Praise for *What the F*

"A skillful presentation . . . *What the F* delivers on the surprise promised by its title, as what seems like a book about language taboos turns out to be a cognitive scientist's sneaky—charming, consistently engrossing—introduction to linguistics. . . . Bergen synthesizes reams of his own and others' research clearly and cracks some pretty decent professorial jokes. . . . Entertaining and enlightening." —*New York Times Book Review*

"A delightful new book. . . . Studying swearing is a way of studying human nature itself." —*Economist*

"Offers useful information." —*New York Review of Books*

"A sweeping book, exploring not just the history of English profanity in words and in gestures, but also the impact that swears and other taboo words can have on the human brain. . . . A valuable addition to the literature about profanity." —**Atlantic.com**

"*What the F* is accessible and engaging, and so brimming with insights that even as a linguist, I found myself stopping every couple of pages to say to myself, 'Huh—I never thought of that.' You'll find yourself saying the same thing—and you'll never hear profanity the same way again." —**Geoff Nunberg, language commentator, NPR's *Fresh Air*, and author of *Ascent of the A-Word***

"In *What the F*, a self-proclaimed 'book-length love letter to profanity,' cognitive scientist Benjamin K. Bergen succeeds in bringing me around to appreciate the broader context, as well as the finer points, of the role 'bad' words play in human society." —*Science*

"An illuminating read, and makes the case for swears as a salutary aspect of our lexicon." —**A.V. Club**

"Full of cute tidbits you can drop at cocktail parties. . . . It's a quick read, not a detailed, academic dissection. But don't mistake breeziness for triviality: cursing plays a central role in our lives." —**Ars Technica**

"Interesting and insightful." —*National Review*

"Some prospective readers may avoid this book because of its subject matter. That would be a gosh-darned shame." —*Science News*

"A fascinating journey to the crossroads of etymology, neuroscience, and culture." —*Discover*

"Bergen has a flair for constructing arguments and discussing complex ideas in accessible language, helped by vivid examples and a splendid sense of humor . . . combine that with reams of fascinating science and analysis and you have an irresistible item for fans of *fuck* and friends." —*Strong Language*

"Oh, it's a lot of fun, and scientifically sound too!" —*Language Hat*

"*What the F* is rigorous enough to guide future scientific inquiry, and casual enough to be read by any ordinary bastard with a passing interest. At the very least, this book reassured me of the profundity of my own human capacity for expression when I rolled out of bed last month to find out who got elected president of the United States and could only utter that one favorite curse word." —*PopMatters*

"There's something here guaranteed to offend everyone (the book wouldn't be doing its job otherwise), but . . . lovers of language will savor every word." —*Booklist*

"A lively study with the potential to offend just about anyone. . . . From a linguistic and sociological viewpoint, the book is illuminating, even playful. . . . An entertaining . . . look at an essential component of language and society." —*Publishers Weekly*

"A winner for the psycholinguistics nerd in the house." —*Kirkus Reviews*

"This book is a surprisingly engaging introduction to a topic rarely discussed or examined. . . . Highly recommended." —*Choice*

"You might think a book about cursing would tell us that lately people seem to be doing it more, that the F-word goes a long way back, and maybe that in the Old West people used to say 'Tarnation!' Benjamin K. Bergen reveals how much more there is to profanity, ranging from why *poo* doesn't end in a consonant through how people curse (or not) in other countries and

about what things, to whether we should formally ban slurs, and on to the Pope and the brain. *What the F* teaches us that profanity is not just pungent, but as *interesting* as other aspects of the miracle we call language."
—John McWhorter, author of *The Power of Babel, Our Magnificent Bastard Tongue,* and *The Language Hoax*

"Profanity is about powerlessness and power. Powerlessness leading to frustration, anger, surprise, and in positive cases, awe—where speech acts are easier, or preferable to, physical acts, or the only possibility. Power where there is intent to harm and words hurt. Fear of the natural emotions behind profanity have led to a need for order and politeness and the view that profanity is 'bad language.' Profanity is natural, and as such, is a lens into emotion, cognition, and cultural norms. It takes courage, energy, extraordinary intellectual chops, and a sense of fun to take on profanity. Benjamin K. Bergen has all in full measure. Read this book." —George Lakoff, professor of cognitive science and linguistics, University of California, Berkeley

"An elegant, insightful, and ballsy application of rigorous linguistic methods to swearing, that most revealing—and ignored—corner of language. Census data reveals where all the Dicks have gone. Brain imaging shows how we manage to avoid taboo slips of the tongue. A careful analysis of studies about profanity's alleged harm to children betrays them as the anti-profanity agitprop they are. Though a descriptivist to the core, I issue the following prescription: read this effing book!" —Jesse Sheidlower, author of *The F-Word*

"Why we swear and where and when it is permissible are explained in this compelling treatise on one of the most taboo subjects in all culture. Read this fucking book or else you might be a wanker." —Michael Shermer, publisher, *Skeptic* magazine, monthly columnist, *Scientific American,* presidential fellow, Chapman University, and author of *The Moral Arc*

WHAT THE F

Also by Benjamin K. Bergen:

Louder Than Words: The New Science of How the Mind Makes Meaning

WHAT THE

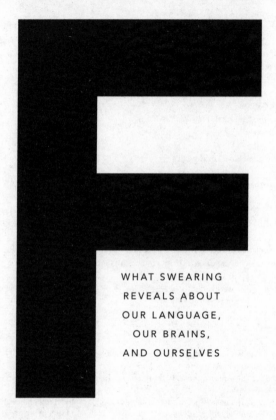

F

WHAT SWEARING
REVEALS ABOUT
OUR LANGUAGE,
OUR BRAINS,
AND OURSELVES

BENJAMIN K. BERGEN

BASIC BOOKS

New York

Basic Books
Hachette Book Group
1290 Avenue of the Americas, New York, NY 10104
www.basicbooks.com

Printed in the United States of America

First Trade Paperback Edition: April 2018

Published by Basic Books, an imprint of Perseus Books, LLC, a subsidiary of Hachette Book Group, Inc. The Basic Books name and logo is a trademark of the Hachette Book Group.

The Hachette Speakers Bureau provides a wide range of authors for speaking events. To find out more, go to www.hachettespeakersbureau.com or call (866) 376-6591.

The publisher is not responsible for websites (or their content) that are not owned by the publisher.

Print book interior design by Jack Lenzo

The Library of Congress has cataloged the hardcover edition as follows:
Names: Bergen, Benjamin K., author.
Title: What the f : what swearing reveals about our language, our brains, and ourselves / Benjamin K. Bergen.
Description: New York : Basic Books, 2016. | Includes bibliographical references and index.
Identifiers: LCCN 2016007705 | ISBN 9780465060917 (hardback) ISBN 9780465096480 (ebook)
Subjects: LCSH: Words, obscene—Psychological aspects. | Swearing—Psychological aspects.
    | Blessing and cursing—Psychological aspects. | Languages, Modern—Obscene words—
    Psychological aspects. | Psychoanalytic interpretation. | Psycholinguistics. | BISAC:
    SCIENCE / History. | LANGUAGE ARTS & DISCIPLINES / Linguistics / General. |
    PSYCHOLOGY / Social Psychology.
Classification: LCC P410.O27 B47 2016 | DDC 401/.9—dc23
LC record available at https://lccn.loc.gov/2016007705

ISBNs: 978-0-465-06091-7 (hardcover), 978-0-465-09648-0 (ebook), 978-1-54161-720-9
    (paperback)

LSC-C

10   9   8   7   6   5   4   3   2

This book is dedicated to my mom and dad. You cultivated my fascination with language and discovery. Also my chutzpah. Thank you for your unconditional love and support. Without it, writing a book like this would be unthinkable. And in order to preserve it, maybe stop reading right now. Things are about to get hairy.

Contents

Introduction

This is a book about bad language. Not the tepid pseudoprofanities like *damn* and *boobs* that punctuate broadcast television. I mean the big hitters. Like *fuck*. And *cunt*. And *nigger*. These words are vulgar. They're shocking. They're offensive. They're hurtful.

But they're also important. These are the words people use to express the strongest human emotions—in moments of anger, of fear, and of passion. They're the words with the greatest capacity to inflict emotional pain and incite violent disagreement. They're the words that provoke the most repressive regulatory reactions from the state in the form of censorship and legislation. In short, bad words are powerful—emotionally, physiologically, psychologically, and socially.

And that makes them worth trying to understand. To someone like me, a cognitive scientist of language and one with a pretty foul mouth at that, profanity is a gold mine. Where do these words come from? Why do we have them at all? What would a world without profanity look like? How do taboos about language vary across the world's languages and how are they similar? What does exposure to profanity do to our brains? What does it do to our children's brains? How do slurs like *nigger* and *faggot* acquire their unequaled capacity to cause harm? What if anything can be done to remedy their impact on marginalized individuals and groups? Can we ban, censor, or reappropriate our way out of harmful words? Addressed with care and attention, these guiding questions can lead toward a cognitive science of profanity.

Bad language deserves inspection on its own merits. But it's also important for a second, perhaps slightly less obvious reason. Profanity is powerful, so it behaves differently from other types of language. It gets encoded differently in the brain. It's learned differently. It's articulated

1

differently. It changes differently over time. And as a result, bad language has the unique potential to reveal facts about our language and ourselves that we'd otherwise never imagine. Studying profanity teaches us where language gets its power to shape minds and to shape the world, how our brains learn language, and how language must have evolved. Throughout its several-thousand-year history, the scientific study of language has, if anything, mostly tried to ignore profanity. But I'm prepared to make the argument that this has been to our disadvantage. In certain ways, you can learn more from four-letter words than from fifty-cent ones.

Perhaps I can make my case a bit clearer with an analogy.

Recently, my wife and I had our first child. I wouldn't call us naive, but leading up to our son's birth, we focused mostly on the positive things that soon-to-be parents often envision. Snuggling with a larval infant, a first smile, sharing giggling sessions, his first step, and so on.

Within minutes of the little guy's arrival, however, we confronted a very different reality. The daily experience of parenthood, at least early on, predominantly entails the monitoring and containment of the child's bodily functions. I'll concede that a baby is by definition a human. But in practice, a baby is functionally a machine for converting milk into bodily effluvia. And an efficient machine at that. As a result, a large proportion of our time was quickly filled with figuring out the best way to clean a rug soaked in baby spit-up. Or a shirt covered in tar-like baby poop. Or a lamp shade drenched in baby urine. You get the idea.

The various substances that emerge from the infant are a nuisance, and they're gross. At least, at first they are. The thing no one tells you about being a parent is that among the things that change (like the diameter of your waist and your tolerance for sleep deprivation) is your relationship to things that come out of another human's body. And like many parents, we came to treat inspections of diaper contents as a diagnostic tool. If you haven't been through this yourself, it might sound strange, but it actually makes a lot of sense. You see, infants are inscrutable. It's hard, for example, to know how much milk a newborn is taking in. (Breasts don't have volume markings on the side, and they aren't transparent. Two more ways evolution has failed us.) But you can tell how much the newborn is nursing from the quantity and frequency of wet and soiled diapers. You even receive a chart at the hospital. In the first week of life, you're told, look for six wet diapers and two dirty ones per day. Or here's another reason to inspect the diaper—one for the real breast-feeding insiders. How do you know if the child is spending long enough on each breast? That's right, it's in the poop. If he's draining each breast, he'll be getting not only the lean fore-milk, which will turn his stool green, but also the fat-rich hind-milk, which will turn it orange or brown. You're hoping to find the latter in the diaper.

And here's the thing. Once you get over the initial aversion, diaper contents actually turn out to be pretty darn interesting. If you love the child and are concerned about his well-being, it follows that you care about what goes in and, as a consequence, how it looks when it comes out. And that's because there are things to learn about your infant by attending to the gross stuff that you just couldn't know if you attended only to the ap-pealing stuff, like smiles and cooing. Changes in his stool may be your first indication that he's ill. And inspecting his vomit might be the only way to know definitively where that set of tiddlywinks went.

While the blissful, sanitized, halcyon ideal of parenthood that many of us begin with might be seductive (and in fact might be necessary to get any of us to willingly commit to it in the first place), the truth is that there is a dirty side too. And that dirty side is, in its own way, beautiful. OK, maybe beautiful goes too far, but at the very least, it's revealing. You learn more about an infant—what he needs, what he's eating, and how he's feeling—by also looking in the diaper. And over time, you come to appreciate it.

$ % !

I only mention all of this because what I've come to learn about babies is also true of language.

The ancient Sanskrit grammarians of the fourth and fifth centuries BC discovered and documented the patterns of sound and meaning that still form the basis for our modern scientific conception of language. And since then, philosophers, linguists, anthropologists, sociologists, and psycholo-gists have studied how language works—how people make new words, how they move their mouths to articulate sounds, and how words change over time. The subject matter is fascinating. Language is fascinating. But over the last twenty-six centuries language scholars have focused on a sanitized and saccharine type of language. For the most part, language scientists have only been talking about the pretty part of the metaphorical baby. And that's a pity, because you actually learn far more by considering the dirty parts too.

Let me give you two examples—two ways in which dirty language re-veals things about language that we'd never have suspected otherwise.

We've known for a long time that specific parts of the brain play special roles in language. The critical bit of evidence is that when these certain parts of the brain suffer damage, due for instance to a stroke, lesion, or trauma, people start to have trouble pronouncing or understanding words. But the same brain damage leaves other cognitive capacities unaffected. This tells us that these particular brain regions are important for language. But there's a twist, and it involves profanity. Damage to language-supporting brain regions doesn't impair all language equally. In fact, a lot of the time, even when brain damage obliterates most language, swearing still remains. And people with brain damage do swear. A lot. (They do have a lot to swear about, what with the brain damage and all.)

This fact usually gets swept under the rug in discussions of language disorders or how the brain encodes language. But it's important because it means that the automatic, reflexive swearing that spurts out when you stub your toe or get cut off on the highway uses different parts of the brain from the rest of language. Language, we've come to find out, isn't all localized in the same place in the brain. The story is far more complex and far more nuanced than that. But we only know this because of the *shits* and *goddamnits* that leap from the mouths of people with brain damage who are otherwise linguistically challenged.

Here's another example. Words change their meaning over time. Sometimes they become more general. For example, in English, the word *dog* actually once referred to a particular kind of pooch, something like a mastiff. Now it's used for dogs in general. It's changed. Conversely, words can get more specific. The English word *hound* used to mean "dog" in general (you might suspect this if you know some German, where the word *Hund* still refers to any dog), but now *hound* refers only to hunting dogs, so it includes greyhounds but not poodles. Fascinating, sure. But why do old meanings go away when words change their meanings? *Dog* and *hound* provide no answer. But there are clues in the dirty underbelly of language. Consider the name *Dick*. I'm willing to bet that you don't know anyone under the age of fifty-five named *Dick*. You know young *Richards* and young *Ricks* but no young *Dicks*. But there are plenty of old *Dicks*. Why? For exactly the reason you think. Once a word gains a new meaning (once *dick* came to refer to the male member), then it becomes problematic to use the same word with its older meaning. The name *Dick* is tarnished by the common noun *dick*. New uses of words push old uses out of the way as a natural part of generational language change. But you wouldn't really understand why words shed their old meanings if you didn't consider where all the *Dicks* have gone.

$ % !

So that's the flavor of what's to follow. The chapters that follow represent deep dives into eleven different dimensions of the science of swearing. Profanity has a lot to teach us about language—not only how it's realized in the brain and how it changes over time but what happens when children learn it, how it hooks into our emotions, and why it occasionally trips us up. But profanity is also fascinating in its own right. We'll investigate where it gets its emotional and social impact, where our beliefs come from about what's appropriate and what's obscene, and how a society establishes and enforces norms for linguistic behavior.

This is an enterprise worth pursuing because despite how prevalent and how powerful profanity is, almost none of us know even the most basic facts about it. Why are the profane words in English profane? Is it something about how they're spelled? How they sound? Where they come from? Are the same words profane across the English-speaking world? How representative is English of the world's languages? What does swearing do to your brain? What does it do in aggregate to a culture with different religious, cultural, or ethnic groups?

Admittedly, we don't have definitive answers to all of these questions. But a few researchers are working on them. These psychologists, linguists, and neuroscientists are not always particularly forthcoming, and with good reason—there are strong taboos at play. Even though many people use profanity, we also tend to think that profanity is not appropriate for certain contexts; indeed, that's what makes it profanity. So if you're a scientist doing research on swearwords or teaching a class on profanity at a public university, you may well experience some pushback—politicians, pundits, or even the public wondering out loud about the value of this particular use of tax money. And while universities are designed to be open forums for intellectual freedom and free speech, that won't keep a professor from being fired for using words like *fuck* and *pussy* in the classroom, as happened to a tenured Louisiana State University professor in June 2015.[1]

So within research institutions, there has long existed an outsized requirement for researchers to justify studying profanity. In linguistics departments, the only day that profanity makes it onto the syllabus is usually when there's no way to avoid it. That's in presenting so-called infixes, such as the *fucking* in *un-fucking-believable*. Because only profane words (or near facsimiles) can be "infixed" into other words in English, linguists feel safe presenting profanity on that day of class. It's the only way to convey the concept. For the most part, though, language researchers steer clear of studying profanity, even if it's potentially fascinating, for fear of what will happen when their institutional review board evaluates their experimental materials or when a committee of their peers reads their publications during tenure deliberations.

Nevertheless, a small cabal of researchers has been toiling away on profanity. With several exceptions, most notably psychologists Timothy Jay[2] and Steven Pinker,[3] they've largely done their work without much public attention. At least until recently, they've been practitioners of a secret science of swearing.

But things have started to change, in large part because of changes in public language norms. The highly regulated public airwaves don't carry the bulk of public communication as they once did. First cable television and then the Internet have created a Wild West for words, where the true will of the people has its way. And if social media are any indication, the people want to be able to swear. And to hear swearing. And to read swearing.

As the public has become more accustomed to profanity, taboo words have started to make their way more prominently into mainstream science. And that's where we are now. And that's why it's time for this book. Profanity needs a little celebration. That's what this book is. It's a coming-out party for the cognitive science of swearing. This is a science that tracks words over centuries as they shift and change, that measures their impact on a child's developing emotional health, and that uses them as a Rosetta stone to the atypical brains of people with Tourette's syndrome or aphasia. At every turn, the dirty, uncomfortable, taboo side of language reveals things that you would never guess if you didn't look. That's what this book is about. It's a guide to what you learn about language when you take a deep breath, hold your nose, and then open up the diaper and take a close look.

1

Holy, Fucking, Shit, Nigger

W ords do things to people. Some words demonstrate such rich erudi-
tion that, when deployed strategically, they cause university professors to swoon. I'm speaking from experience here. Words like *prolixity.* Or
eponymous. Delightful. Let me get started writing your grad school acceptance letter right now. Other words affect people because they're so fleeting—
words like *normcore* or *ratchet* or *on fleek.* Deploy these words at precisely
the right moment in their penetration of the lexicon and you're the coolest
hipster at the indie cold-brewed coffee co-op. But wait a week and you'll be
served your coffee with an eye-roll. Words, in short, have the power, by their
mere utterance, to affect how people feel and how they feel about you.

And the most potent words of all—the ones that have a direct line to
the emotions—are profanity. Profane words uniquely allow you to express
pain or cause it in others. They peerlessly demonstrate frustration, anger,
or emphasis. But let's be specific. I mean words like *cocksucker.* Or *fuck.* Or
cunt. These are among the taboo words of English that elicit the strongest
measurable physiological reactions—the fastest pulse, the sweatiest palms,
the shallowest breathing. These words are versatile. Name a feeling, and
profanity can elicit it. Profanity can increase sexual arousal. It can increase
your ability to withstand pain (compare the analgesic effect of yelling *fuck!*
when you hammer your thumb with the effect of yelling *duck!*).[1] When
deployed appropriately, profanity can cause delight—countless comedians
stake their professional lives on the impact of "working blue." But when
miscalibrated, use of the very same words can make you seem crude, un-
educated, or out of control. In their darkest incarnation, profane words can

be part of verbal abuse, they can denigrate and disempower people, and they can be used in maledictions.[2]

And because these words have such outsized impact, we ban them. We chastise or spank children for using them and fine or arrest adults who use them around children. Because the words are just too powerful.

What gives these words such an intensity and such a diversity of power? Where do they emerge from? Do they work the same in every language around the globe? Or could a language exist without profanity? And what would that look like?

Before looking for ways to answer these questions, we need to define some terms. I'm using the words *profanity*, *cursing*, and *swearing* interchangeably. I realize that some etymological hairsplitters out there will want to distinguish among them. It's true that *profanity* once referred only to blasphemous language (*profane* contrasts with *sacred*). But I'm using it as it's used in modern parlance—where it includes not just religious language, like *Jesus Christ*, but the whole taboo gamut: *fuck, shit, cunt*, and the lot. Following the lead of Timothy Jay in his influential book *Why We Curse*,[3] I'll be using the words *cursing* and *swearing* in the same way. And the same goes for *expletive*. It's not that I don't think there are important differences among the various types of taboo language—quite to the contrary! It's just that for all intents and purposes, people at present don't systematically distinguish what the words *profanity*, *swearing*, and *cursing* refer to.

That said, we do need ways to talk about the various specific types of profanity, or things related to profanity. One is a *slur*, a derogatory term for a person or group of people. These are also called *epithets*, *terms of abuse*, *terms of disparagement*, *derogatory terms*, or *pejorative terms*. People don't always agree on which words are slurs and which are not (and, as we'll see later, it changes over time), but some clear examples in contemporary English are slurs like *nigger*, *faggot*, and *bitch*. Now, not everyone agrees that slurs are profanity—for some people, *nigger* is a swearword, whereas for others it falls into a distinct category of taboo word. In order not to get hung up on definitional issues like this, in this book we'll be considering slurs alongside the more traditional types of profanity and identifying differences where they bubble up from the data. And there are differences. Notably, as we'll see in a later chapter, slurs have the greatest potential to cause harm and therefore demand different treatment.

I snuck one final concept in there that we should probably be clear about: *taboos*. Taboos are social customs—norms or mores—that prohibit certain types of behavior. For instance, there are things you know you're not supposed to do in public—we wall off bedrooms inside houses and toilets inside stalls to satisfy and perpetuate taboos about bedroom and bathroom activities. We often also find it taboo to merely talk about those same things in public.[4] We have taboos about telling people about excretory functions or sexual exploits, for instance. It would violate your expectations about normative social behavior if you asked a job candidate what other things you should know about her and she started telling you about recent abnormalities in her defecation schedule.

Profane words are those particular words that some people in a culture believe are unacceptable in specific settings. The taboo is about the words themselves, not necessarily what they denote. The taboo against the word *shit* is about the word itself; the word is taboo regardless of whether it's used to describe feces or to express frustration. And we know that profanity is about the word rather than the content because in many situations it's perfectly acceptable to talk about the same content using different words. Parents will willingly talk to small children about their *poo-poo* or to their doctor about their *stool*. But if they hear the word *shit* on the radio while the kids are listening, you can bet they'll be sending the station manager an angry letter. And actually, to refine our definition of profanity just a bit further, it's not really the words themselves that aren't acceptable but the words used with specific senses or meanings. Words like *ass*, *cock*, and *bitch* can be passable when used to describe animals but are profane when describing people or body parts.

Now that the stage is set, let's begin to look empirically at profane words. First, how can we tell which are the profane words in a language? Second, how similar is profanity in the languages of the world? Does it draw from the same sources? Are there languages without profanity at all? And third, when languages differ in how they treat profanity, does that tell us anything about the cultures in which they're embedded?

$ % !

Finding out what words the people who speak a particular language think are profane is not a particularly challenging operation in principle. You just have to ask them. But for the most part, there's been very little effort to do so systematically. Even for English, the world's most studied language, and even in the United States, the world's biggest economy and a country that seems particularly invested in regulating profanity, almost no one has bothered to systematically pose the question, what are the profane words in your language?

Even the people who really should ask—because regulating profanity is part of their job—haven't done so. The Federal Communications Commission (FCC), for instance, is responsible for overseeing all television and radio broadcasts transmitted over public airwaves. The FCC regularly issues fines or other sanctions over incidents of profanity. For instance, it fined Fox for the 2003 broadcast of the Billboard Music Awards. In it, Nicole Richie spontaneously and rhetorically mused about the reality show *The Simple Life*, "Why do they even call it *The Simple Life*? Have you ever tried to get cow shit out of a Prada purse? It's not so fucking simple."

Given how firm the FCC's penalties are (it's taken its right to enact them to the Supreme Court on a number of occasions) and how clear it is about the times when children might be listening (precisely between the hours of 6 a.m. and 10 p.m.), it would be reasonable to assume that the FCC has a published list of words you can use during daytime hours, or more to the point, words you can't. But I challenge you to try to find the official FCC list of banned words. Go ahead—fire up your Google box. You won't find it. Because there is no official list. Profanity is something the FCC apparently knows when it sees it—it says profanity is "language so grossly offensive to members of the public who actually hear it as to amount to a nuisance."[5] If you were to give the FCC the benefit of the doubt, you might suppose that despite having no published official list of offending words, still it must have done some empirical research, asking normal Americans how they react to the words in question. But there's no evidence of this. As far as anyone can tell, the FCC hasn't actually done the legwork to find out what people really think about words—what's profane, in the present culture, at the present time. And in the unlikely chance that it has, it's certainly not advertising as much.

This stands in contrast with regulatory bodies in other parts of the English-speaking world, which have actually tried to get an objective handle on profanity. The standard is set by the New Zealand Broadcasting Standards Authority (NZBSA), which is roughly analogous to the FCC. The NZBSA conducts a survey about every five years to see what Kiwis think about a variety of potentially objectionable words and publishes a complete accounting of its methodology and findings.[6] In the most recent round, it asked 1,500 adults to rate how acceptable or unacceptable they'd

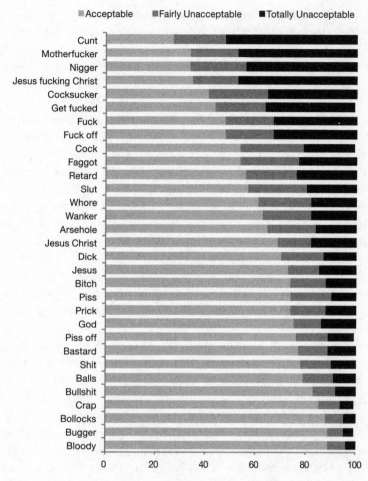

The least acceptable words in New Zealand.

find dozens of words and expressions, should they appear on nighttime television. Without further ado, here are the most unacceptable words in New Zealand, from worst to slightly less worst. The chart above shows the proportion of people who judged each word as falling into one of three categories. The light bars on the left indicate how often each word was judged either "Totally Acceptable" or "Neither Acceptable or Unacceptable." The medium-hued bars in the center reflect the proportion who responded that each word was "Fairly Unacceptable." The dark bars indicate those who found a given word "Totally Unacceptable."

All told, there were only eight words that more than half of survey respondents thought were fairly or totally unacceptable on air after 8:30 p.m.—the ones at the top, from *cunt*, *motherfucker*, and *nigger* down through *fuck off*. Words farther down the list were rated as more acceptable. For instance, only 30 percent of respondents felt that *dick* would be unacceptable in that context, and just 22 percent objected to *shit*.

A couple of things stand out from this list. First, for each word below the top eight, the majority of respondents actually thought that it was acceptable on television. That includes not just *dick* and *shit* but *cock* and *faggot*. Kiwis have a comparatively high threshold for profanity on television, at least compared with what the FCC appears to assume about Americans. Second, and perhaps so obviously that it goes without saying, the survey displays very clearly that people disagree about how unacceptable words are. Respondents were about evenly split on whether *fuck* is acceptable on television or not. Even *cunt* and *nigger* elicited 27 and 34 percent of respondents, respectively, saying that the words aren't objectionable on television. This diversity of opinion prompts a host of second-order questions. How can there be so much disagreement about what's acceptable? Do differences in opinion correlate with other variables—for instance, do opinions about words correlate with ethnicity, gender, age, geography, and so on? These aren't just scientific questions—the same issues confront you if you're in the broadcast standards business. How much agreement do you need on a word to ban it? In a hypothetical case, suppose there's a word that a minority subgroup of the population finds profane, and say it's a term of abuse, like *nigger*. In such a case, which matters more, the opinions of the population in aggregate or those of people in the relevant subgroup? How do you decide?

The New Zealand study can't answer these questions. But we can start to get a feel for how opinions about language differ around the globe by looking elsewhere in the Anglophone world. How does the New Zealand list compare, say, with swearing in the birthplace of Shakespeare? There's no precise analog to the New Zealand survey, but the Broadcasting Standards Commission of Great Britain did release a study in 2000, which it authored jointly with several other groups.[7] The study asked 1,033 adults a series of questions about profanity, including whether each of a list of words was "Not Swearing," "Quite Mild," "Fairly Severe," or "Very Severe." The results are shown on the following page.

The two surveys are hard to compare. For one thing, they asked different questions—about the acceptability of words on television at a particular

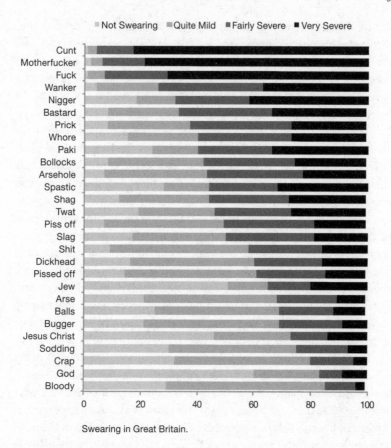

Swearing in Great Britain.

time on the one hand versus the severity of the words in general on the other. In addition, the sets of words they asked people about weren't identical. The slur *Paki* (denoting people of South Asian decent) was judged either fairly severe or very severe by 60 percent of British respondents, but it simply wasn't provided on the New Zealand list. Conversely, *Jesus fucking Christ*, the fourth-least-acceptable word in the New Zealand survey, didn't appear in the British one.

As a result, the absolute ratings of words aren't really worth perseverating on. But the general trends are still informative. Namely, like the New Zealand study, the British study shows rampant disagreement. Half of respondents said that *slag* (a derogatory term for a promiscuous woman, similar to the American word *slut*) was either fairly severe or very severe, while the other half judged it quite mild or not swearing. Second, some words behave similarly across the surveys—for instance, *cunt*, *motherfucker*, and *nigger* show up in the top five in each list, while in both lists *bloody* and *crap* appear not to widely offend. But at the same time, there appear to be substantial regional differences. *Wanker* shows up at a prominent number four in the Great Britain list, just ahead of *nigger*, but on the New Zealand list, it falls in the middle of the pack at number fourteen, right before *whore*. Similarly, *bollocks* finds itself on the more severe side of the Great Britain study, wedged between *Paki* and *arsehole*, but on the New Zealand list it falls nearly to the bottom, coming right after *crap*. So to the extent that these differences are the product of more than the survey instrument itself, they point to potential regional differences in how these words are viewed.

There's been no study of the same magnitude or with the same weight of the state behind it in the United States or Canada. But where regulatory bodies in North America have shied away from profanity, fortunately, academics have stepped up in a smaller way, proportionate to their more modest means. How does North American English compare? Are certain components of the list similar (omitting dialect-specific terms, like *wanker*, *bollocks*, and *get fucked*, that Americans don't typically use)? I know of only two pertinent recent studies. One small and one a bit larger. Let's start with the small one. A couple of years ago, two undergraduates in my lab conducted a survey to get an idea of how profane people think specific words are.[8] To reiterate, it was small—much, much smaller than the ones from New Zealand and Great Britain. We asked twenty native speakers of American English to rate the offensiveness of words from 1 (least offensive) to 7 (most offensive). They appear on the next page, with the most offensive again at the top.

Despite the diminutive nature of this survey, you find some alignment with the New Zealand and Great Britain studies. The top performers in terms of offensiveness should be familiar, with *cunt*, *motherfucker*, and *fuck* near the top of the list. As you continue down from there, you find that respondents really weren't overwhelmingly offended by other words that we usually think of as taboo. For instance, notice that *asshole*, *piss*, and *tit* are actually rated less offensive than *scum*—at least by the people who took the survey. One thing to note is that we didn't include slurs in our little study, because they weren't relevant to the specific purpose we intended to use it for. So we have no information about where slurs would fall.

Fortunately, there's another, larger, better study of profanity in American English. Cognitive psychologist Kristin Janschewitz asked eighty

Mean Offensiveness Rating

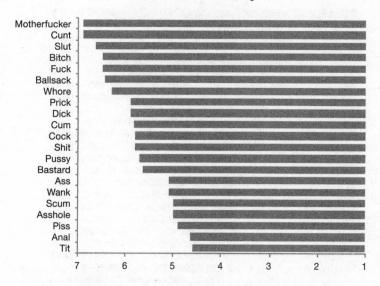

Offensiveness of American English words, small study.

people to rate hundreds of words along a number of axes—not only offensiveness but also how taboo they thought other people thought the words
were, how much they themselves used and were exposed to the words, and
so on.[9] This is a rich resource, and I'll be mining it in the chapters that follow. Janschewitz included ninety-two words that could plausibly be considered taboo. What you see on the next page is how offensive each was rated,
again from most to least profane. I've only included those terms rated most
offensive, down through *shit*. That makes forty-one of them.

Looking at English as it manifests itself across the world, we see hints
of both consistency and variability. Certain words are repeat offenders.
Others are culture- or dialect-specific. But even when the specific words
change somewhat from list to list, the substitutions seem to fit the same
pattern. In Great Britain, *Paki*, *slag*, and *bollocks* all make the list. And if
you know that these are, respectively, a slur for people from South Asia,
a word roughly equivalent to *slut* in the United States, and a word meaning "testicles," it might not be surprising to see them here. At least across
the Anglophone world, central tendencies capture the types of words that
people find offensive, unacceptable, or profane. Profanity isn't random. It's
principled. Let's articulate the principle.

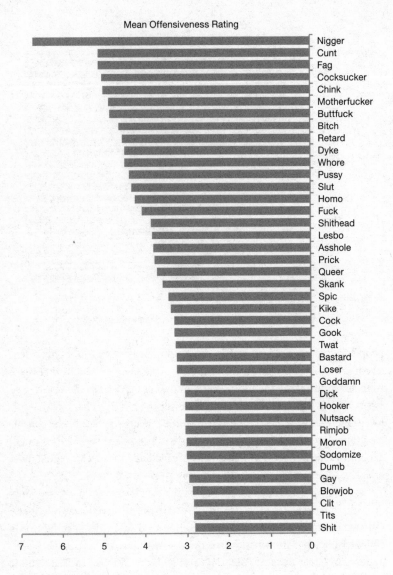

Offensiveness of American English words, larger study.

$ % !

English profanity tends to be drawn from certain categories of words.

The word *profanity* originally referred to the first group. In Latin *profanus* literally means "outside the temple," denoting words or acts that desecrate the holy. For some people, the use of religious words in secular ways constitutes blasphemy—a sin against religious doctrine—and this is the pathway that makes those terms taboo. The names of religious figures, like *Jesus Christ, Jehovah,* or *Mohammad,* are easy fodder. So are aspects of religious dogma. In English, we have a few of these, like *holy, hell, God, damn,* and, of course, *goddamn.* There are also older English curse words that have fallen out of favor, like *zounds,* which derives from *God's wounds,* and *gadzooks,* from *God's eyes.* When terms like these are removed from their sacred context and stripped of their religious intent—when they're "taken in vain"—they can function as profanity. The line between desecration of sacred concepts and profanity is subtle, and as we'll see later, historical religiosity is one of the best predictors that a language will have a robust system of profanity. But for the present purpose, we need only note that the first place English profanity originates is the sacred.

The second place English profanity comes from is language relating to sex and sexual acts. This includes the acts themselves (*fuck,* for instance), sex organs involved in those acts (*pussy* and *cock*), people who perform those acts (*cocksucker* and *motherfucker*), and artifacts and effluvia related to those acts (*spooge, dildo,* and so on). So the second prong of our profanity principle is sex.

Third is language involving other bodily functions—things that come out of your body, the process of getting them out of your body, and the parts of your body that they come out of. This includes robust cohorts of words describing feces, urine, and vomit, among others, as well, of course, as the body parts associated with these substances and the artifacts used in those body parts' upkeep, like *douchebag,* and so on.

And finally there are the slurs. Among the most offensive words on each of the lists (when the lists saw fit to ask about them) are terms like *nigger, faggot, retard,* and the like. These words are offensive by dint of their derogatory reference to people based on some group that they're perceived as belonging to, defined in terms of sex, sexual orientation, ethnicity, religion, and so on. New terms like this are developing all the time—relatively recent additions to English include *tard* (from *retard*) and *sperg* (derived from *Asperger's syndrome*).

Looking just at English, you'll find that nearly all the most profane words in Great Britain, New Zealand, and the United States fall into one of these four categories: praying, fornicating, excreting, and slurring. This is an important point, important enough to name a principle for it. I hereby propose we call it the Holy, Fucking, Shit, Nigger Principle.

Many of the most offensive words on the four surveys fall into the Fucking group. A *wanker* is one who masturbates. *Cunt* refers to a Fucking-related body part. And, of course, many of the words actually have the word *fuck* in them. The tops of the lists are also populated by *nigger* and other slurs. Lower on the lists are Shit-category words, words related to bodily effluvia, like *shit* itself, *asshole*, *piss*, *puke*, and so on. They're not as vulgar, but they're still on the list. Holy-category words, at least in English, seem relatively tame.

How generalizable is this pattern? If it captures something about human nature or about the inevitable evolution of cultural systems, then you'd expect it to apply broadly. Across the world, the vast majority of taboo language should be drawn from one of these four domains, perhaps even in similar proportions. Alternatively, English speakers might be a breed apart, uniquely obsessed with religion, copulation, bodily functions, and social groups. If you pick your favorite language other than English, how does profanity work? What's profane in Cantonese? How about Finnish? Does the Holy, Fucking, Shit, Nigger Principle stand up?

$ % !

Systematic research on profanity in English may be sparse, but there's enough of it to go on. Other languages have basically zilch—no large-scale surveys and no small ones either. So if you want to know what profanity looks like in, say, French or Japanese, you have to dig around through language guides for foreigners with a particularly saucy bent, the occasional academic paper, interviews with native speakers of the various languages, or the rare regulatory document describing what words are banned where and by whom. These kinds of sources are limited in that they all encode the opinions of one or a few people—they're not the product of systematically collecting data from native speakers of the language. But that's what we have to go on, and that's what I've relied on to produce the following assessment of how well the Holy, Fucking, Shit, Nigger Principle does around the world: pretty well.

Cantonese has five words widely agreed upon as the most vulgar in the language—these are the words censored on broadcast television in Hong Kong.[10] They are *diu* ("fuck"), *gau* ("cock"), *lan* ("dick"), *tsat* ("boner"), and *hai* ("cunt"). If you're keeping score at home, that's five for Fucking.

Or consider Russian. Ripped from the censor's press sheet is the official list of the most profane Russian words, currently banned from movies, plays, and other forms of art.[11] The strongest profanity in the language, known as *mat'*, has two tiers. The top tier houses the four most profane Russian words: two words for genitalia, a word equivalent to *fuck*, and a word that translates as *whore*. Including the second tier of somewhat outdated and weaker profanities, *mat'* totals eleven words: seven for genitalia, plus two for sexual acts and two for categories of people who engage in stigmatized sexual acts (prostitutes and homosexuals). In sum, two slurs and the rest are related to sex.

Finnish, which is unrelated to Russian and Cantonese (or to English for that matter) paints a similar picture, at least based on accounts provided by linguists. The top Finnish profanities are words roughly equivalent to *hell*, *God*, *cunt*, *piss*, *shit*, *ass*, *fuck*, and a number of words roughly translated as *cunt* or *cock*.[12]

And so it goes in language after language. Most of the profane vocabulary in most languages that have accessible documentation is drawn from one of these four categories. That's not to say there aren't local exceptions. One is language about animals—calling someone a dog in Korea is deeply offensive, for example. Disease often creeps into profanity, and a salient example is Dutch, which counts among its strong profanities words for cancer, typhoid, and tuberculosis.[13] Ostensibly, in Dutch, the severity of the illness communicates the strength of the profanity. Another rare but attested source is words derived from maledictions—literal curses, like *Damn you to hell!* Or *A plague on both your houses!* And there are taboos about death and death-related words. For example, across many cultures, there's a taboo against naming the dead. Once a person dies, his or her name cannot be uttered, sometimes for a year or longer, as in some Australian Aboriginal cultures,[14] and sometimes under penalty as severe as death, as among the Goajiro of Colombia.[15] But these are fluid. For the most part, when a language and culture designate a stable set of words as profane—where the words themselves are deemed inappropriate and offensive—these largely follow the Holy, Fucking, Shit, Nigger Principle.

Curiously, not all languages hew to the principle in the same proportions. Languages almost always have a mixed portfolio of swearing drawn from the four pillars, but they also invest unevenly in them. Some languages draw so disproportionately from religious terms to populate their profane lexicons that you might want to call them Holy languages. By the same reasoning, there could also be Fucking, Shit, and even Nigger languages.

By this measure, Quebecois French is a Holy language. It makes heavy use of what it calls *sacres* ("consecrations")—strong profanities related to Catholicism and Catholic liturgical concepts. Far stronger than *merde* ("shit") or *foutre* ("fuck") in Quebec are *tabarnack* ("tabernacle"), *calisse* ("chalice"), and *calvaire* ("Calvary"). This is despite—or due to—the fact that Quebeckers have largely lost their religion. The "Quiet Revolution" of the 1960s left most of them Roman Catholic in name only. And yet the holy curses persist, even in the face of a populace that has lost touch with the sacred origins of the words.

And Quebecois isn't the only Holy language. Italian has a set of words similar to the Quebecois *sacres*, known as *bestemmie*. Most involve adding the word *porco* ("pig") to words for Catholic figures, like *porco Dio* ("pig God") or *porca Madonna* ("pig Madonna"). Similarly, in some dialects of Spanish, *ostia* ("host") is profane, as is naming the virgin (*La Virgen*) or the "blessed chalice" (*Copón bendito*). It's no coincidence that these are languages spoken in places where the Roman Catholic Church has had a significant cultural presence. And while Catholics don't have the market cornered on Holy-derived profanity, they nevertheless are laudably consistent in populating local profanity with religious terminology.

Fucking-category languages are more pervasive. A good example is Cantonese, which, as I mentioned earlier, uses words for the act of copulation like *diu* ("fuck") or relevant body parts, like *tsat* ("boner"), as its strongest terms. Same with most varieties of English—as we saw earlier, whether in the United States, New Zealand, or Great Britain, the majority of the words judged most profane or most inappropriate relate to sexual acts, the organs used to perform them, or the people who engage in them. By this measure, Hebrew is probably also a Fucking language, although due to its unique history (the language had largely died out and was reconstructed in its modern form in around 1900, predominantly from religious texts), most of its swearing is borrowed from other languages, like English and Arabic. And Russian is quite clearly a Fucking language, with all of its *mat'* referring to sexual organs, acts, or actors. Not a hint of Holy or Shit.

Shit-category languages are harder to come by. There's a case to be made for German; although some strong profanity in German is drawn from the Holy and Fucking domains, it's not as pervasive as in English. The German equivalent of *fuck*, which is *ficken*, is not commonly used in swearing. But German has a lot of Shit talk. Some of the most used and likely most familiar expressions make use of or are built from *Arsch* ("ass") and *Scheisse* ("shit"): *Arschloch* ("asshole"), *Arschgeburt* ("born from an asshole"), *Arschgesicht* ("ass face"), *Sheisskopf* ("shithead"), and so on.

The similarity of these examples to English might tempt you to say that English is in fact something of a Shit language as well. After all, the Anglophone swearing quiver is full of *shit-* and *ass-*related words. Consider *dumb-shit, shit-faced, shit-balls, shit-sticks, shit-sack, shit-canned, shit-fit, shit-house, shit-load, asshole, ass-face, dumb-ass, smart-ass, ass-eyes, ass-clown, ass-hat*, and I could go on. English is full of *shit*.

So we've seen plausible examples of Holy, Fucking, and Shit languages. Are there Nigger languages? Perhaps English. Among the words that many native speakers consider worst are ones originally drawn from derogatory terms for individuals or groups with certain attributes. *Nigger* might be the strongest modern example, but in my classroom at the University of California, San Diego, many students feel similarly about *chink* and *beaner*. English has profane terms based not only on ethnicity but also on sex (*bitch, cunt*), sexual orientation (*fag, dyke*), immigration status (*wop, FOB*), and health condition (*retarded, sperg, lame*). (See, for a historical perspective, the delightful book *Holy Sh*t*.[16]) But these terms are largely limited in their use. In English, *fuck* is everywhere. So is *shit*. You don't have to be talking about copulation or defecation for these words to find a niche. But that's not quite as true about *nigger* and *chink*. These are strong words, but they haven't migrated as robustly away from their sources.

Of course, classifying languages into one bin or the other isn't particularly important, and it serves to gloss over the subtleties—most languages draw from a variety of sources for profanity, and many profane terms blend together words with diverse pedigrees, like *Jesus motherfucking Christ* or *holy fucking shit*. So let's not get lost in the weeds.

Ultimately, I want to highlight just two things. First, languages tend to draw from similar domains for their profanity. The Holy, Fucking, Shit, Nigger Principle isn't just about English. It's about language. And that suggests that the forces that make words become profane in English may be present across human experience, regardless of native language. And

second, despite similarities across languages, cultural idiosyncrasies play a
role in shaping how profanity in a language will work and how it will be distributed. Languages spoken by people with a cultural history of uniform religious practice for instance (read here: Catholicism) can become populated
with Holy profanity—words for heaven and hell and saints and demons.
You might even say that when it comes to cultural differences in profanity,
the *devil* is in the details. (Or you might think better of saying that.)

$ % !

As far as cultural differences go, tilted distributions of Holy, Fucking, Shit,
and Nigger words are only the beginning. We Anglophones have a regulatory bent. Many of us feel that certain words by their very nature are bad
and potentially harmful. And our impulse is to regulate language, through
rules like laws that limit free speech in order to maintain profanity-free
airwaves, movie theaters, and public spaces. But this is not a human universal. Just casting a glance around the society of nations reveals stark cultural differences in the suppression of profanity.

In France, for instance, even the most profane words of the language,
like *foutre* ("fuck") and *putain* ("whore") are so common that if no one
told you they were bad words, or *gros mots* ("fat words") as they're called in
French, you could be excused for not figuring it out yourself. There's no con-
certed censorship of specific words in the media in France as there is in the
United States, which is part of the reason these words are everywhere. These
words also have many distinct uses, which vary in terms of their strength
and meaning. For example, *foutre* ("fuck") is used as a general verb meaning
something like "do" or "give." For instance, *Qu'est-ce que tu a foutu?* literally
translates as "What is it that you fucked?" but its meaning is more like "What
the fuck did you do?" To say that someone has an estimable physique, you
can say that he or she is *bien-foutu*, literally "well-fucked" but more equivalent in English to "well-fucking-built." The same versatility is true of *putain*
("whore"), which is used a lot like English *fuck* as a general intensifier. It can
go at the beginning of a sentence: *Putain, ça coute chère!* ("Whore, that's
expensive!") which means something like "Fuck, that's expensive." Or *J'en
ai marre de cette putain de voiture!* ("I'm fed up with this whore of a car!"),
which would be used equivalently to "I'm fed up with this fucking car!"

There are certainly limits in France to how widely these terms can be
used. But nearly everyone uses them, from television personalities to the

prime minister.[17] While you might do well to avoid using them in your first interview with a potential employer, they're certainly less offensive to French people than lots of other things you could say. It's not that France is a paradise for linguistic libertarians. Clearly some utterances are inappropriate, such as verbal abuse or solicitation of undesired sexual interactions. But profanity isn't as taboo in France as it is in, say, the United States.

Cultural attitudes toward swearing can be even more foreign. In some cases, a language can be totally bereft of profanity. Consider the curious case of the missing Japanese profanity. You might be familiar with the contours of the story from the James Bond book and movie *You Only Live Twice*. At one point, Bond is in Japan, training with Tiger Tanaka, head of the Japanese Secret Service. Bond casually swears, and Tanaka reacts with a short comparative linguistics lesson, explaining that, "There are no swear-words in the Japanese language and the usage of bad language does not exist." According to Tanaka, in those moments of heightened and transient motion that elicit expletives from an Englishman, a Japanese speaker would only utter things like *shimatta*, "I have made a mistake" or *baka-yaro*, "fool."

Although the Bond oeuvre isn't necessarily renowned for its anthropological sensitivity and nuance, in this particular case, Tiger Tanaka's story stands up pretty well to scrutiny. Japanese does have specific ways of speaking that are thought to be stronger than others, and there are many ways to insult people. Beyond uttering potentially insulting words like *bakayaro*, you can offend someone by using the wrong grammatical form of a verb or noun—similar to how an English speaker might offend his surgeon by addressing her as Carla instead of Dr. Lee. Japanese even has a special way you're supposed to talk to the emperor, with its own prescribed noun and verb forms, without which you could surely offend.

But as Tanaka says, Japanese seems to largely lack a core feature of what makes the profanity we're familiar with in the English-speaking world so complicated and so powerful. Profane English words like *fuck* aren't proscribed just because they insult people or because they describe sexual acts. There's something about the words themselves that we consider bad. And this key element appears not to be a cultural universal. In Japanese, you can insult people directly by calling them names. And Japanese has words for genitalia and for acts of deploying them. But there's reportedly no real equivalent to the class of English words we consider profanity; nor is there any societal agreement that those words are "bad" and need to be regulated.

Being curse-less has consequences. It affects the things you can do with the language—the work you can do with words. So Japanese speakers who want to swear have to look elsewhere. Take Ichiro Suzuki, a Japanese baseball player who spent a large part of his career playing in the United States for the Mariners and Yankees. Ichiro is a polyglot—he speaks English when appropriate but also uses Spanish with players from Latin America and the Caribbean. He told the *Wall Street Journal* in an interview, "We don't really have curse words in Japanese, so I like the fact that the Western languages allow me to say things that I otherwise can't."[18] If you want to curse in Japanese, you literally have to do it in English or Spanish.

On the opposite end of the spectrum from Japan are societies in which some agency is authorized to regulate and restrict public language use. To some extent, the United States is such a place in that there are certain exceptions to the right to freedom of speech, and one of those is profanity. (Much more about that in Chapters 9, 10, and 11!) But there are far more authoritarian language regimes to be found. During the writing of this book, for instance, Russia banned a list of profane words from the arts—books, theater, films, music, everywhere. Violators will be fined. The particular words targeted are unsurprising—they're the most profane words that I mentioned earlier, *mat'*. Those words are *khuy* ("cock"), *pizdá* ("cunt"), *yebát'* ("to fuck"), and *blyad* ("whore"). (It now occurs to me that these developments might make this book hard to purchase in Russia.) And it's not just these four words that are now banned but all words that include them. You see, Russian, like English, likes to build off of its profane words, which makes for a rather lengthy list of banned words. For instance, *pizdá* ("cunt") can be augmented in a variety of ways: *pizdéts* is used as an exclamation, meaning something like "deep shit!" The verb *pizdét'* means "to lie," a close equivalent to English "bullshit." And so on.

If you keep looking, you can find more repressive regimes. There are places where use of taboo language, in particular blasphemy, is treated as a capital crime. For instance, countries or parts of countries governed under strict sharia law (from Afghanistan to Yemen) punish blasphemy with death. This is a way of taking prohibitions on taboo language to the most violent extreme—if using words in certain ways is bad, and if it's the state's (or the church's) responsibility to act in the interest of the well-being of individuals, then it follows that the state ought to use its punitive apparatus to impose limitations on speech.

So although languages tend to draw from similar sources to populate their lexicons of profanity, those commonalities are eclipsed by cultural differences in what people think about words. A culture appears to be able to decide whether or not to buy into the idea that certain words deserve to be called out for special treatment. A language doesn't have to have profane words. And that's a point worth remembering when we return later to the question of censorship and the future of profanity.

$ % !

In what ways are the 7,000 languages of the world similar, and in what ways are they different? Both questions have fascinated linguists and philosophers for millennia, for different reasons. Universal features found to hold in all languages reveal something about what it is to be human. If all humans do something—whether it's art, music, math, or some aspect of language, that universal behavior must be due to either some shared common experience or some trait possessed by all humans, transcending cultural idiosyncrasies. Perhaps, sometimes, this stems from our genetic endowment.

There doesn't appear to be much about profanity that is truly universal— shared without exception by all languages and cultures. It's not just that the specific words are different. As we've seen, the differences are much deeper than that. Some cultures have rich and deeply codified systems of profanity, like English or Russian. Others, like Japanese, don't really have anything like the same category of words. Instead of absolute universals, when we look around the globe we find certain common tendencies across languages. The Holy, Fucking, Shit, Nigger Principle takes a first stab at characterizing the types of words that tend to become profane. Languages select from a small pool of semantically constrained candidates for their bad words—if indeed they decide to have bad words. Not only do the specific words differ from language to language, but so do the proportions of words selected from each domain, in ways related to the sociocultural legacy that a given language carries with it.

But this sort of statistical universal, where features overlap in languages that exhibit family resemblances, is the norm in the languages of the world—not just when it comes to profanity but for language features in general. It's very hard to find much of anything that all languages do. When you look for universal features of languages, you mostly find tendencies. This makes us think that the way a language will be structured isn't merely random. Something must be at work making languages similar, but it isn't some inviolable rule inscribed in our genes. In each case of a cross-linguistic tendency, facts about how people use language—what they want to convey with it, the memory and time constraints imposed on them while using it, and so on—likely shape languages over the course of generations such that they settle on certain similar sorts of solutions. For example, people seem to want to talk about things and events, so it's not surprising to find nouns and verbs in the world's languages. Similarly, it can be useful to distinguish who did something from whom they did it to. As a result, languages evolve subjects and objects and ways to encode them. So if profanity is like other cross-linguistic tendencies—languages tend to have it, and it tends to be drawn from certain domains—then what pressures tending to produce similar-seeming profanity could the histories of the world's languages share?

The answer probably lies in taboos not about language but about the world. Across cultures, people exhibit taboos about the very things that provide the vocabulary for profanity. There are taboos around the world associated with the supernatural—with gods and demons and prophets. There are taboos about copulation. There are taboos about defecation, micturition, menstruation, and other bodily functions. And there are taboos about people who are not members of our social group (see, for instance, laws against miscegenation that remained on the books in the United States until 1967!).

The fact that taboos like these erupt around the world, though not universally, suggests an explanation for how profanity comes about and how it comes to have similar contours. People around the world have taken these taboos and extended them from the world to the word. It's not just defecation that's taboo in many cultures; nor is it just talking about defecation. Rather, the words that describe defecation themselves are taboo, whether that's how you happen to be using them in the moment or not.

There could be different reasons for this. We know that merely hearing or seeing a word stokes an internal mental representation of the things the word refers to.[19] If the word *shit* causes people to "see" feces in their mind's eye and "smell" it in their mind's nose, then the impulse to limit the word's use is understandable. Or it could be that people hold more metaphysical beliefs about words and their power—that they believe that using words associated with a particular taboo topic will bring bad fortune.

Whichever of these explanations is ultimately correct—and there's more work to do to tease them apart—the specific words that are profane across languages are similar because the things that are taboo across cultures are also similar. The pressure to reject words associated with those taboos is the real universal.

But here's the catch. The road from taboo things in the world to taboo words is nondeterministic. Even if excretion is culturally taboo, that doesn't mean that all words describing it will be as well. *Shit* is more profane than *poop*. *Fuck* is profane, but *copulate* is not. And so cultural taboos only set the stage for profanity. They don't select specific words. What distinguishes profane *cunt* from childlike *wa-wa*? That's up next.

2

What Makes a
Four-Letter Word?

Across the globe, profanity tends to emerge from particular domains of meaning—I refer you to the Holy, Fucking, Shit, Nigger Principle. But for every profane *holy*, *fucking*, and *shit*, there's a technical and anodyne *liturgical*, *copulation*, or *excretion*. For every *cock* and *cunt*, there's a childlike *wee-wee* and *cha-cha*. Many words describing sexual organs, excretory functions, and so on fail to rise to the heights (or, if you prefer, sink to the depths) of profanity. These words are articulated without fear of offending, whether in the classroom or the courtroom or the examination room. They aren't profane, despite referring to taboo concepts. This means that something beyond what a word denotes—what it refers to—must cement it as profanity.

What is that thing?

Why is *cunt* a dirty word when *coochie-snorcher* isn't?

The most obvious possibility is that some aspect of how profane words are written or sound makes them vulgar. Let's begin with the eight-hundred-pound gorilla. Many English profane words famously have four letters—not just *cunt* but *fuck*, *shit*, *piss*, *cock*, *tits*, and many others. No matter how you count, a lot of the profane words in English are spelled with four letters. Take just the words from the four lists in the last chapter. These lists aren't exhaustive. But what's nice about them is that they weren't assembled with any particular interest in what the words sound like or how they're spelled. Admittedly, the people who had to come up with lists of profane words might have been unconsciously swayed by the four-letter word notion, but at least that wasn't their stated objective. So in that way, they offer as unbiased a sample as we're likely to find. Those four lists in aggregate give us a total of eighty-four distinct words (I've removed multi-word expressions like *get fucked* and *Jesus fucking Christ*, which include other words already in the list). Of the eighty-four words, twenty-nine are spelled with four letters. By this count, then, just over a third of profane words are four-letter words. This number may be artificially deflated, since many of the longer words (like *asshole*, *motherfucker*, and *wanker*) have shorter four-letter words embedded inside them. But it's a good start.

The first thing to notice from this is that having four letters isn't a necessary prerequisite for profanity. Certainly, we already knew this: words like *ass* and *motherfucker* don't have four letters, and most of the words on the list have some number of letters other than four. Nor is having four letters sufficient, since many four-letter words are not at all profane, like *four* or *word*. So we have to reconsider the question we're asking. The real issue seems to be whether having four letters makes a word more likely to be profane, all other things being equal. That's still an interesting question. Here's a way to ask it.

Given that many (but not all) profane words in English are spelled with four letters, we can try to find out whether the pattern is stronger than you'd expect, given how words in the language are spelled generally. That is, suppose you grabbed a random set of eighty-four English words. What are the odds that twenty-nine of them would have four letters? You can see a histogram on the next page, showing how many profane words from our list have each number of letters—the profane words are in the dark bars. As you can see, there's a sharp spike at four, representing those twenty-nine four-letter profane words. But is twenty-nine a lot? You can tell by comparing the lengths of profane words in dark bars with English words in general, shown in the light bars. (To calculate these values, I counted the English words with each number of letters and normalized these counts to an eighty-four-word language to make them directly comparable to the profane numbers).* As you can see, English has a lot of words with four, five, six, or seven letters. And in general English looks like a smoother version of the profane distribution. But what really sticks out is how many more profane four-letter words there are than expected from English in

---

* To calculate the numbers for English in general, I used the lemmatized word list that Adam Kilgarriff generated from the British National Corpus (available from his web page).

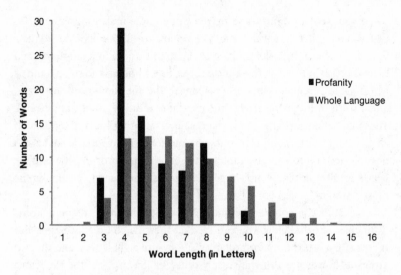

Profane words (dark bars) are more likely to be three, four, five, or eight letters long than are English words in general (light bars).

general. The 29 profane four-letter words in our list are significantly more than you'd expect if profane words were like English words in general, in which case we'd expect only 12.6 profane four-letter words out of 84.*

Perhaps more surprising is how many profane three- and five-letter words there are. There are relatively few three-letter words in English overall, and profane words are almost twice as likely to have three letters than you'd expect, all things being equal. We'll come back to this in a moment, because it's important. Less important but also notable is the little bump in eight-letter profane words, compared with the language in general. This is due to words composed of two four-letter words, like *ballsack*, *bullshit*, *buttfuck*, and *shithead*. Four-letter words appear to bend how English words look even when they're merely parts of other words. But for our present purposes, it's enough to note that profanity in English is strikingly more likely to have four letters than other words. The take-away is that there's some truth to the popular notion about four-letter words.

So this raises the obvious question, why? Why are profane words more likely than other words to have four letters?

---

* A chi-squared test of lengths three through twelve reveals that the two samples are significantly different. For the statistically minded, $\chi2(3) = 38.61$, $p < 0.0001$.

If you were a linguist, and maybe you are, the first thing to occur to you would be that the special length of profane words might be due to their frequency of use. In general, the most common words in a language tend to be shorter (in English, these include *the, be, of, and, a,* and so on), and as words get less frequent, they also get longer (the one-thousandth most frequent word in English is *useful,* the five-thousandth is *gravity,* and so on). The explanation for why this is the case is fascinating (having to do with efficiency of information transmission), but for our purposes it could also possibly account for the aberrant lengths of profane words. Maybe profane words are shorter than words in general because they tend to be among the most common.

In fact, if you compare profane words with the most frequent words in English, shown below, you can see that they match up a lot better. But there's still a little bump for profanity at four and eight letters, and the two groups of words are still statistically different.* So this can't be the whole explanation, but it might be part of one.

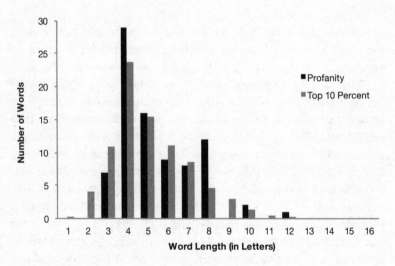

Profane words are also more likely to be four or eight letters long than the most frequent 10 percent of English words.

---

* I ran a two-by-ten chi-squared test for lengths one to ten comparing profane words with the 626 most frequent words from the British National Corpus: $\chi2(9) = 19.17$, $p < 0.05$.

The catch is that it's hard to know whether profanity really is as frequent as the top 10 percent of English words. The difficulty lies in the fact that sources we usually use to measure word frequency are all written, mostly in a formal language register—things like newspaper archives and great literature. Profanity is vanishingly rare there. But the informal and spoken environments that form its natural habitat and in which it thrives leave no record. So we can't measure how common it is in those places. Here's the best I can do. I searched in a place where people do use language relatively casually and that does leave a permanent, searchable record: the website Reddit. Reddit is an interactive news, entertainment, and commentary platform fractured into various topic-related communities. People can post links or comments and often interact with informal language. They also tend to be younger than the population at large and more male. I took the eighty-four profane words in question and computed how frequent they were on all of Reddit, on average, over the course of two years, from August 31, 2013, to August 31, 2015. Profane words were quite frequent—not quite as frequent as those in the top 10 percent, but close.

So the upshot is that frequency might explain part of why profane words tend to be four letters long in English. But it doesn't tell the whole story. Perhaps there's something else going on—perhaps something about having this number of letters causes a word to seem particularly taboo. Indeed, in some places in the world, people avoid the number four systematically—you can think of it as the number thirteen of Southeast Asia. More on that later, but the association between four letters and profanity appears to largely be an English-specific phenomenon. Although we can't do comparable analyses for other languages (because lists of profane words in other languages haven't been systematically tested), a quick tour around the swearwords of the world reveals that the four-letter rule doesn't apply in many other languages. Often the most profane words in non-English languages are a different length. For instance, the strongest French words, *putain* ("whore") and *foutre* ("fuck"), have six letters, and there is almost no Mexican Spanish four-letter profanity—strong words are longer, like *chingar* ("fuck"), *concha* ("cunt"), and *pinche* ("fucking"). In some other languages, profane words aren't spelled with four letters because there's no spelling at all—in places where tetraphobia (fear of four) is pervasive, the local languages often aren't spelled with an alphabet. Chinese, for instance, uses logographic characters instead. And more generally, spelling is only relevant to the half of the world's languages that have a written form.[1] So if spelling is responsible for the four-letter phenomenon, then it would have to be for English-specific reasons.

And when you sit with the idea for a bit, other considerations might cause you to second-guess whether having four letters could really make words profane. After all, people have been speaking English for a thousand years, and for most of that time many of those people couldn't read or write. But they could swear. Children can swear before they can read or write (more on that later!). And even within English, some pretty profane words happen not to have four letters but are pretty close, like *ass* or *bitch*. So perhaps we're detecting obliquely, through spelling, another, deeper cause at play. Maybe profane words tend to be four letters long because four-letter words tend to be pronounced a particular way. Maybe *shit, cunt, fuck,* and the like don't look profane so much as sound profane.

$ % !

This might seem outlandish, but hear me out. Take a moment to think about profane four-letter words, like *cunt, fuck,* and *shit.* Doesn't something about them just sound dirty? Don't they sound vulgar? Don't they sound aggressive?

If you agree, you're not the first person to intuit that the words of your language somehow sound appropriate for what they mean. This was noted at least as early as the nineteenth century by German linguist Georg von der Gabelentz,[2] who observed that German speakers consider the French silly for calling a horse *cheval* rather than the clearly more suitable German word *Pferd.*[*] Truly, though, *cheval*? Ridiculous! Obviously it's a *Pferd.* Even if you know intellectually that there are different names for horses in other languages, in your heart of hearts, you may still feel like *horse* in your native language fits the animal best and that equivalent words in other languages are slightly less apt. This is sometimes called the "sound-symbolic feeling": the sounds of words in your language feel like they suit their meanings.

Taboo words often elicit particularly strong sound-symbolic feelings. When you say them—*fuck, shit, bitch*—when you roll them around in your mouth, they have a certain mouth-feel. And gut-feel. They feel like they sound obscene. One manifestation of this feeling is that it's hard to

---

[*] In case you're wondering, yes, it's really pronounced with a *p* followed by an *f*!

imagine them meaning anything else. How could *fuck* signify anything other than what it does? (For instance, a word that sounds very similar in French, *phoque*, preposterously means "seal.") And so we're baffled when people who are not native speakers of our language accidentally produce profane words. Spanish speakers often confound English *sheet* and *shit* or *beach* and *bitch* because Spanish doesn't encode a distinction between the *ee* and *i* sounds. Even knowing this, the sound-symbolic feeling makes it almost inconceivable to a monolingual English speaker that you could think that *sheet* would be pronounced *shit*. *Shit* feels dirty. *Sheet* shouldn't.

So could something about how profane words sound make them profane? Does this sound-symbolic feeling index something more than a subjective feeling? Do *shit* and *fuck* sound objectively more vulgar than *poo-poo* and *copulate*?

One of the most common reasons words sound appropriate to their meanings is that the things they refer to sound like something, and the word's pronunciation reflects that sound. This familiar phenomenon is known as onomatopoeia or sound symbolism. Words for sounds or actions that produce sounds often (but not always) imitate those sounds themselves. For example, even if you didn't know what they meant, with a little context, you might be able to make an educated guess about the meanings of *cock-a-doodle-doo* and *swish*.

Could profanity be sound symbolic? There are some good candidates; consider words like *barf* or *piss*. Of course, the word *barf* isn't a perfect imitation of what vomiting sounds like; nor is *piss* an exact replication of the sound of micturition. Still, there's enough of a resemblance between the words and their referents to create a semblance of sound symbolism.

But how can we tell? This is a hard problem because we don't have a great way to measure sound symbolism. One brute-force approach would be to just ask people to report how sound symbolic a word seems to them, say, on a scale from one to seven. Researchers do this a lot. But this really only tells us what words English speakers subjectively think are sound symbolic—it's an index of their sound-symbolic feeling. We, however, are looking for an external, objective measure of whether the words would sound like what they mean even if you weren't already a speaker of the language.

So a slightly more nuanced way to measure symbolism is to take a list of words from one particular language, say, English, and present them to people who don't speak that language, say, monolingual Japanese speakers. And then you ask them to do something like guess the meanings of the English words. You have to do a lot to set up an experiment like this right. You have to use participants who really haven't been exposed to any English and English words that haven't been borrowed into Japanese. You have to be sure that there aren't any similar Japanese words just by chance. But if you get it all right, then in principle the English words guessed more easily by people who speak only Japanese (or any other non-English language) are more likely to be sound symbolic.

But to my knowledge, no one has ever done this systematically with taboo words. So we don't know. And in any case, it's unlikely to work. For one thing, the implementational details would make it hard to pull off. For instance, it's getting harder and harder to find people around the globe who aren't familiar with some English, especially profanity. So you'd probably encounter the most success if you used profanity from Finnish or Basque or some language with a lower profile than English. But the deeper issue is that it's unlikely in principle that sound symbolism of the *cock-a-doodle-doo* type is in play for a lot of profane language. Sound symbolism is most common and most effective with words that describe either sounds or things that systematically make stereotypical sounds. *Barf* has potential for sound symbolism because it describes an action that makes a canonical and recognizable sound. Same with *piss*. But there are few other viable profane candidates: maybe *crap*, *queef*, and a couple others. Most profane words are ill-suited to sound symbolism because the things they refer to don't systematically sound like anything. *Bitch* isn't a good candidate, at least not in the taboo use, because there's no sound associated with a malicious or unpleasant person that can be imitated. And the same is true for language about sacred concepts (what does God's prophet sound like?) and sexual organs (what does a penis sound like?).

But the real death knell for profanity being generally sound symbolic comes when you compare profane words with similar meanings. If words sound like what they mean, then words with similar meanings should also have similar sounds. For instance, there's reason to believe that *moan*, *groan*, and *whine* are sound symbolic not just because they individually sound like the sounds they denote but because they have both similar meanings and similar sounds. Likewise, if *fuck* somehow sounds like what it means, then other words with similar meanings should sound similar. But they don't; a good comparison list might include verbs like *bang*, *bone*, *dick*, *shag*, *screw*, and so on. Consider how these words sound and how they're spelled. Most don't share any sounds at all.

And you get the same insight when you compare words across languages. Look at words that are the translation equivalents of *fuck* in other languages. French has *baiser*, Spanish *chingar*, Mandarin *cào*, Russian *yebát*, and so on. Even at first glance, and even including in our little sample only languages that are very closely related and that maintain close cultural contact with one another, despite their similar meanings, these words sound nothing alike. None of the sounds in *fuck* are in any of these other words (the *c* in the transliteration of *cào* is pronounced more like English *ts* than *k*). The words are different lengths, they contain different sounds, and they're written differently. And the same is true of *shit* and *bitch* and any profanity you want to try out. These words, across languages, behave more like *horse*, in that the various words don't share a resemblance, than *neigh*, where they do.

The point is that no matter how apt *fuck* feels to express the concept it does, when you turn to the next language, people have the same feeling about their words—French *baiser* or Spanish *chingar*—which use totally different sounds. By this measure, the sounds used to express the meanings of these words appear arbitrary. That is, it appears that nothing about the sounds in the word *fuck* makes them particularly apt to express the meaning of the word *fuck*. And nothing makes the sounds of *fuck* a better fit for its meaning than the sounds of *cào*. Over the course of the history of English, French, Spanish, Chinese, and thousands of languages on earth, words have evolved that do similar social work—that fit a similar communicative niche—but are pronounced in very different ways.

The upshot is that while some profanity might sound the way it does because of sound symbolism, this is unlikely to be true for the majority of taboo words. (At least not in the words of spoken languages. But in the next chapter we'll look at gestures and at the signs of signed languages, where the story is revealingly different.) Perhaps the sound-symbolic feeling we get with profanity is really just the result of a lifetime of using a word with a particular sound to mean a particular thing. If seeing a horse (or smelling or hearing or feeling it) often goes with hearing or thinking or saying the word *horse*, then why wouldn't you develop a strong association between the sound and the meaning, especially if that's the only language you know? And likewise for profanity.

So it seems that sound symbolism isn't what makes profane words sound dirty.

$ % !

If not sound symbolism, perhaps some other aspect of how profane words sound makes them seem dirty. Let's loop back around to where we started. English exhibits a higher proportion of four-letter words among profanity than among words in general. As you'll recall, there are also more profane three-letter words. So let's dig into these words. What's special about how these three- and four-letter words sound?

The three-letter words included in the list are *ass, cum, fag, gay, god, Jew,* and *tit.* And the four-letter words are *anal, anus, arse, clit, cock, crap, cunt, dick, dumb, dyke, fuck, gook, homo, jerk, jism, jugs, kike, Paki, piss, scum, shag, shit, slag, slut, spic, suck, turd, twat,* and *wank.*

Do you notice any general trend in how these words are pronounced?

Here's one idea. I haven't seen it discussed anywhere in the literature on profanity before, but it occurs to me that if you look closely, the three- and four-letter words tend to have two properties. First, regardless of how many letters they're spelled with, they tend to be pronounced with just one syllable. In case you need a refresher, a syllable is a rhythmic beat of language, during which the mouth opens and closes. When you pronounce *bitch* and *shit* normally, they're only one syllable long.* Just a few words on the list have more than one syllable: *anal, anus, homo, Paki,* and, arguably, *jism.*

Now, this can't possibly be the whole story, because there are thousands of one-syllable words in English, and most of them aren't taboo—11,752 to be precise (with the vague notion of precision appropriate for counting words in a language).† The profane words are but a speck in a sea of monosyllables. And if we're just looking at three- and four-letter words, it's no surprise that they'll tend to be pronounced with one syllable or two.

But these words don't just tend to be monosyllabic. They tend to be built in a particular way. English allows many different types of syllable. Every syllable has a vowel at its core.‡ For some syllables, the vowel is both the beginning and the end (the alpha and the omega, as it were), as in words like *a, I,* and *uh.* (Don't be confused by spelling—there's no *h* in the pronunciation

---

* That said, you can opt to make them into two-syllable words, as in *Sheeyit, what a gigantic beeyotch!* And if you're not into the whole brevity thing, you can even turn it up to three syllables with *biz-nee-atch* and *shiz-nee-at.*

† The count of all monosyllabic words appearing in the MRC Psycholinguistic Database at least once is 11,752: Wilson, M. D. (1988).

‡ Or something vowel-like. A word like *hurdle* has two syllables, but neither has an easily recognizable vowel. Yet both the *ur* and the *le* can anchor syllables.

of *uh*.*) But most syllables also have consonants in them, before or after the vowel. So with this in mind, we can return to English profanity. If you briefly revisit the words in the lists above, you may notice something remarkable about their syllables. I'll wait for you to discover it yourself.

Got it? I'll give you a hint. There are four exceptions. They're *gay*, *Jew*, *homo*, and *Paki*.

Here it is. Every other word on those lists ends with one or more consonants. That is, they all have "closed syllables" rather than syllables sporting bare vowels. (A decent mnemonic is that your mouth closes at the end of a closed syllable.) As you can see, many profane words even double down on their final consonants. Words like *cunt* and *wank* actually have two consonant sounds at the end. Interestingly, consonants seem pretty important in general—all but a few (like *ass* or *arse*) begin with at least one consonant, and many begin with two, like *crap*, *prick*, *slut*, and *twat*. But really the strong generalization here appears to be that syllables of profane words tend to be closed.

Could these two tendencies—a trend toward having just one syllable and another toward that one syllable being closed—be part of what makes profane words sound profane?

We can start to answer this by splitting our data in a different way—based not on how many letters a word is spelled with but on how many syllables it has and whether those syllables are closed. When we do that, we find that not just the three- and four-letter words are closed monosyllables; so are seven of the sixteen five-letter words, like *balls*, *bitch*, *prick*, and *whore*, but not *Jesus* or *pussy*. In all, thirty-eight of the eighty-four words on the list are one syllable long, and thirty-six of these (or 95 percent) are closed. Only two profane words on the list, *Jew* and *gay*, are "open" monosyllables (the *w* and *y* aren't pronounced as separate consonants—they're part of the respective vowels they follow). How does this ratio compare to the words of English more generally? I took the top 10 percent most frequent monosyllabic words from the MRC Psycholinguistic Database, which has both frequency information and phonetic transcriptions for English words. It turns out that whereas 95 percent of our profane monosyllabic words are closed syllables, that number drops down to 81 percent when you look at nonprofane words, which is significantly lower.[†]

---

* Unless you're Butthead.
† Fisher's exact test: $p < 0.05$.

You might now be scrambling to find exceptions—profane monosyllabic words in English that are open. Our list of eighty-four words definitely doesn't cover all profane words in the language—we were using it, as you recall, because it was constructed without any explicit prior expectations about the sounds or spellings of profane words. And you can probably find some profane open monosyllables. Like, potentially, *ho, lay, poo,* and *spoo.* These are good candidates. Maybe you can come up with one or two more. But for each one, there are a dozen closed monosyllable candidates that we left out of our initial list. They include, in alphabetical order, *boob, bung, butt, chink, cooch, coon, damn, dong, douche, dump, felch, FOB, gook, gyp, hebe, hell, jap, jeez, jizz, knob, mick, MILF, mong, muff, nads, nards, nip, poon, poop, pube, pud, puke, puss, queef, quim, schlong, slant, slope, smeg, snatch, spank, spooge, spunk, taint, tard, THOT, toss, twink, vag, wang,* and *wop.* And I'm only getting started. Run the numbers again with these new open and closed monosyllabic words, and you still have upward of nine out of ten profane monosyllables that are closed.

This pattern is statistically real, but we really want to know whether it's psychologically real too. Do English speakers think that closed monosyllables sound more profane than open monosyllables? There are different ways to figure this out. Here's one type of circumstantial evidence. When English speakers invent new, fictional swearwords, do they tend to be closed? For instance, when English-speaking fantasy and science fiction writers invent new profanity in imaginary languages, what do those words sound like? *Battlestar Galactica* has *frak* ("fuck"). *Farscape* has *frell* (also "fuck"). *Mork & Mindy* had *shazbot* (a generic expletive). Dothraki, the invented language in HBO's *Game of Thrones,* has *govak* ("fucker") and *graddakh* ("shit"). Not all are monosyllabic, but they all end with closed syllables. In fact, it's very hard to find fictional profanity ending with open syllables. The one glaring counterexample I've been able to dig up comes from the movie *Star Wars: Episode 1,* in which *poodoo* means "bantha fodder" and is used as a weak expletive. Just by way of speculation, the open syllable might have been selected because the target audience of the movie appears to have been quite young (it was rated PG), and so a more profane-sounding fictional profanity could have felt too strong.

We can also indirectly assess the psychological reality of profane closed syllables by looking at real words that are not taboo by dint of their meaning but happen to have closed syllables. Do people think of these words as obscene despite their innocent meanings? In fact there's a phenomenon known as word aversion, in which some people have particularly strong reactions to particular words, even though the words have totally anodyne (or inoffensive) meanings. The English word that appears to crawl most insidiously under people's skin is *moist*. I can't tell you how often, upon discovering that I'm interested in profanity, people declare their everlasting hatred for this word. I suspect that the fact that *moist* is a closed monosyllabic word has something to do with it (along with aspects of its meaning). But to date I know of only one piece of empirical research on word aversions,[3] and it focuses exclusively on *moist*, so if there are indeed other words that people find to be the linguistic equivalent of nails on a chalkboard, it's impossible to know what those words sound like.

But alien languages and word aversions really only supply very indirect evidence about profanity. The best way to tell whether people feel that closed monosyllables are more profane than open monosyllables would be to conduct a study with invented English words, ones that differ only in what kind of syllable they have. You could ask people how profane those words would be if they were real English words. Would people feel that *cheem* is more vulgar than *chee*? Is *smoob* more profane than *smoo*? That way, you could control for all other differences between the closed and open monosyllables and measure whether having a final consonant alone is enough to push the profanity needle.

So I ran this study. I generated a bunch of potential monosyllabic words of English that happen not to be real English words, like *chee* and *smoo*, and I paired up each open monosyllable with a closed monosyllable that was identical except for the last sound. So *skoo* went with *skoom*, and *stee* was paired with *steesh*, and so on for twenty pairs of words that were the same on all the relevant dimensions and different only in the type of syllable.* I also manipulated how many consonants there were at the beginning of the word, known as the "onset," just to see if this also made a difference in how profane the words sounded to people. So of the twenty pairs of words I created, ten began with just one consonant, like *dee* and *deeve*, and the other ten pairs began with two consonants, like *smee* and *smeef*, always an *s* followed by some other consonant, because that happens to be a way English likes to put multiple consonants at the beginning of a syllable.

---

* Open and closed monosyllabic words weren't significantly different in neighborhood density, mean positional phoneme probability, or mean biphone probability, none of which you would ever have heard of if you weren't a psycholinguist but all of which you would be very concerned about if you were.

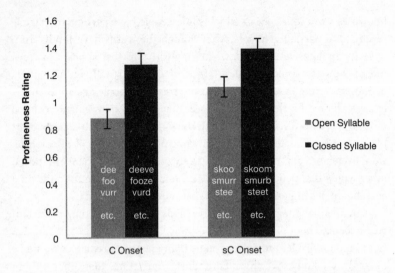

People rate made-up words as more profane when they have more consonants, either at the beginning of the syllable or at the end.

And then I asked sixty native speakers of English, "How profane does the following made-up English word sound?" on a four-point scale from "Very Profane" to "Not at all Profane." You can see what they thought in the graph above. (Words that start with just one consonant are shown under "C onset," and those with an *s* and another consonant are "sC onset.")

Pretty clearly, when everything else is held constant, native English speakers think that closed syllables sound more profane than open syllables (the dark bars are higher than the light ones). Also of interest, and slightly more surprising, there appears to be a weaker though significant effect where having more consonants at the beginning of the word also makes a word seem more profane (the pair of bars on the right is higher than the pair on the left).

So not only does English profanity tend to be pronounced with closed monosyllables, but English speakers moreover think that closed monosyllables sound more profane than open ones. In terms of how languages work in general, this isn't entirely unprecedented. Sometimes within a language, you will find clusters of words with similar meanings that happen to have similar forms. These arise not because their forms reflect their meanings through sound symbolism but for another reason. Consider words in English that have meanings related to light or vision. Many of them happen to start with *gl.* I'll give you a few: *glisten, glitter, gleam, glow, glare, glint.* And there are many more, from *glaucoma* to *glower.* Now, it's impossible for sound symbolism to be at play here because light and vision don't sound like anything at all, and even if they did, there's no reason to think it would be *gl.* Instead, we've uncovered a little dense spot in the English lexicon where words with similar meanings have similar forms for no better reason than that they do.

The story of how these sets of similar words come about goes something like this. In general, sound symbolism notwithstanding, words arbitrarily pair together forms and meanings. But because the words of any language are governed in part by chance, there will happen to be some places in the lexicon of a language where a couple words that have similar meanings happen also to have similar forms. People who learn and use this language may notice these little clusters, or they may not (for example, you may or may not have noticed English *gl*-words before), but over time the clusters will act as a form of attractor for new words. Old words that are misheard, mislearned, or misremembered will be slightly more likely to gravitate toward the form and meaning of a cluster, which appears to have happened in the history of the *gl*-words in English. And new words that people invent will also be attracted to the clusters such that they're slightly more likely than chance to have meanings and forms aligned with the growing pattern. This, too, has happened in the history of English: see examples like *glitzy* (in 1966) and *glost* (a glaze used in pottery, in 1875).[4] It's also a factor in product naming—imagine which glass-cleaning spray you'd prefer to buy: Brisserex or Glisserex. Over centuries, maybe even millennia, these clusters are reinforced in a kind of rich-get-richer process until you have English, where a healthy 39 percent of words starting with *gl* relate to light or vision.[5]

And perhaps this is what happened with English profanity. Perhaps through historical accident there came to be a core set of profane English words that happen to be pronounced with a closed monosyllable. They exerted a gravitational tug on words around them—existing words came to be pronounced similarly, and newly coined words were more likely to follow the same pattern. We can see this in our newest profanity, where acronyms like *MILF, THOT,* and *FOB* tend to be closed monosyllables. And we can see it in the profane abbreviations that people have created over the years, like *gyp, hebe,* and *smeg.*

$ % !

We began by asking if something about words like *fuck* and *cunt*, aside from their meaning, makes them profane. By following the four-letter road, we discovered a hidden pattern in how profane words sound in English. At their core are closed monosyllables. This isn't just a descriptive fact about the words that are currently profane in English; it also affects what English speakers think about new words, whether inventing a science fiction language or participating in a behavioral experiment.

To reiterate, though, there are many exceptions to this closed monosyllable pattern. Not only are there a few profane words with open monosyllables, like *gay, spoo*, and so on, but there are also many profane words that have more than one syllable, like *asshole, motherfucker, cocksucker*, and company. But this shouldn't be too surprising to the well-weathered linguist. Languages exhibit few exceptionless rules. We all know that English makes past tense forms of verbs by adding *-ed*. Except that it often doesn't, in so-called irregular verbs like *spend, go*, and *drink*. English nouns place stress on the first syllable and verbs on the second (compare *a record* and *to record, a permit* and *to permit*). But then sometimes they don't—*copy* and *double* are pronounced with first-syllable stress as both verbs and nouns. So it's no surprise that we can't find a hard-and-fast rule about how English profane words sound. As with these other generalizations about language, there's a tendency. Just as English profanity tends to be drawn from certain semantic domains, so it tends to sound a certain way.

This trend and the fact that it has exceptions might explain differences among words with similar meanings. Words like *poo, pee, gay, Jew*, and *spoo* are all arguably profane words. But if the closed monosyllable pattern is real inside the heads of English speakers, then all other things being equal, words like these should seem less profane than words with similar meanings that are pronounced with closed syllables. Indeed, when you contrast them with closed versions, they might seem to have less oomph. Which is more profane: *pee* or *piss*? Compare *spoo* with *spooge*. *Jew* with *hebe*. *Gay* with *fag*. Does it seem to you like the closed-syllable words are somehow more profane? If so, how well they fit with the closed monosyllable pattern might be responsible. And it might also predict how well they maintain their oomph over time and how widely they're used. As a closed monosyllable, *spooge* ought to end up more widely disseminated as a profane word than would an open monosyllable like *spoo*.

And, of course, the polysyllabic profane words in English still have to be considered. In a way these words are exceptions to the closed syllable trend, and in another way they aren't. More than half of the polysyllabic words on our profane list (twenty-seven of forty-six) begin with a profane closed monosyllable, like the *cock* in *cocksucker* and *wank* in *wanker*. And even more of these same words (thirty in total) end with a closed monosyllable, like *bastard* and *faggot*. The numbers become a little muddier when we try to count these composed words—we could consider dozens that include *shit*, *fuck*, *dick*, or *cum* in them, and we'd have to make arbitrary choices about what to count. But even without going there, we see clearly that English profanity is built in part from closed syllables, whether by themselves or as part of longer words.

If this closed monosyllable pattern is real, where does it come from? I offered an analogy with *gl*-words earlier, suggesting that there doesn't have to be an intrinsic motivation in terms of what the word means and why a particular sound would be well suited to it. For a cluster to take off, it need only be sufficiently frequent. Perhaps, somehow in the history of English, the ratio of open to closed monosyllables in English shifted locally in the subclass of profanity. And that little tilt in the lexicon snowballed.

In keeping with this story, the closed monosyllable principle isn't a cross-linguistic universal. Some languages don't allow for anything like the range of closed syllables we have in English. For instance, syllables in the Hawaiian language can never end with a consonant—they're always open. So there's no possible Hawaiian version of the English closed syllable pattern. And many of the most profane words you might now be familiar with from other languages are open syllables or polysyllabic: French *putain* ("whore"), Spanish *chingar* ("fuck"), Russian *yebát'* ("fuck"), and so on. But as I've mentioned, we don't have reliable studies of profanity for most languages. As a consequence, it's hard to know whether the English pattern shows up in other languages as well.

I want to raise the possibility of another explanation for why profane words in English sound the way they do. It's possible that some of those sounds are particularly well suited to the functions that profanity serves. To be clear, I'm not talking about sound symbolism. It's not that the words might sound like what they mean. The idea instead is that they might sound the way they do because that way of sounding is effective for the way you want to use the words.

This could work in several different ways, in principle. One way is based on the difference in childishness of open and closed monosyllables in English. It just so happens that as they're learning a language, children are first able to pronounce open syllables. That's why a child typically says *ma* and *mama* before *mom*; a child substitutes *ba* for *ball* and *da* for *that*.[6] (We'll have a lot more to say about this in Chapter 8, when we explore where children's little potty mouths come from.) As the child's motor system matures, she then develops the capacity to coordinate consonants not just at the beginnings but also at the ends of syllables. So on the basis of those developmental facts, here's a just-so story. Maybe open syllables sound more childlike because they are in fact easier for children to pronounce. Perhaps people unconsciously associate words that are harder to pronounce with the adults whose motor systems can in fact articulate them. So closed syllables—and words with lots of consonants at both ends—sound like words adults but not small children say.

If this story is true, then we'd expect profanity to show a preference for not only syllable types that are harder for children (closed ones) but also sounds that are harder for them. We'd expect to see sounds like *th*, which is hard for kids, rather than *p*, which is easier. And we'd expect profanity to eschew the repetition of syllables (known as reduplication) that's typical of infant and toddler speech: *mama*, *baba*, and so on. Something like *poo-poo* would be the epitome of a childlike and therefore nonprofane word.

That's one possible foundation for the cluster of profane words we see in English. Here's another, equally speculative explanation. Perhaps short, closed words are more useful than others for swearing. There's an argument to be made that monosyllabicity is useful for expletives—when you slam your finger in a car door, you don't exactly have a lot of time to express what you're feeling. Short words are simpler and more direct. That's the monosyllabic part. Now to the consonants. Perhaps having a consonant at the end works particularly well for words, like profanity, that are deemed inappropriate in some settings. An open syllable just keeps going, whereas having a consonant at the end seals the word in silence. This is especially visible in epithets or slurs, derogatory labels for groups of individuals, which overwhelmingly follow the pattern (think of *hebe*, *chink*, *gook*, *jap*, *WOP*, and so on). These are precisely the type of word you might want to be able to cut short and mumble into your beard. A closed syllable permits that.

We can actually test this seemingly far-fetched idea by looking at precisely what types of consonants bring up the rear of English closed monosyllables. The key is that not all consonants are created equal. Some consonants bring a word to an immediate halt—notably consonants known

All English Monosyllables   Profane English Monosyllables

 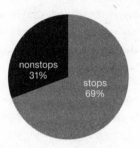

Profane English monosyllables are significantly more likely to end with a stop consonant, like *t* or *k*, than other English words.

as "stops" or "plosives," like the sounds behind *p, t, k, b, d,* and *g*. Other consonants allow sound to continue being emitted—you can prolong a nasal *n* or *m*, a fricative *s* or *f*, or an approximant *l* or *r*. English monosyllabic words in general show a healthy preference for stop consonants over others in their final position—just under half of them, as you can see above, end with a brief, percussive sound, like *p, t,* or *k*. But split profane words in the same way and, as you see above, the bias toward stop consonants is significantly more exaggerated. There are far more profane words like *spic* and *twat* and far fewer like *piss* and *cum* than we'd expect by chance.* This is far from conclusive evidence, but it does lend a little credence to the "shut your mouth" explanation for profanity's tendency to end with a consonant, and not just any consonant.

It's possible that one or a combination of these pressures has chiseled the cluster of profanity that we now see in English. But it could alternatively just be a matter of historical accident, like the case of *gl*. Without systematic studies across languages, we may have to settle for merely observing the pattern of profanity pronunciation in English, in combination with the kind of idle speculation that the last several paragraphs have illustrated.

But one avenue of human communication—a way in which we communicate profanity—has, unlike words, very clear motivation. Beyond words, we also use our bodies to communicate—articulating with our

---

* Specifically, a Fisher's exact test reveals that profane words ending with stop consonants are significantly more frequent than would be expected from the lexicon in general, p < 0.01.

arms and hands, orienting our torsos, and shifting our eyes. We do so
both in the everyday gestures that accompany or replace speech and also,
among people who are deaf or hard of hearing, through the signs of signed
languages. And in the hands, as opposed to the mouth, it's much clearer
why the signals we send—including the obscene ones—have the forms that
they do.

3

One Finger Is Worth a
Thousand Words

Sometime when you're in public—in a park or a restaurant—take a good look at humans and how they communicate. To do this right, you need to suspend what you already know, or think you know, so it helps to imagine yourself as someone with absolutely no prior expectations. Someone like an anthropologist from Mars.[1] Pretend that you're here to study the humans, and just watch what they do to communicate. As a Martian anthropologist, you will surely note how much flailing about there is of parts of the body that contribute strictly nothing to the sounds of the words. Fists shake. Heads cock. Shoulders shrug.

The visible body, deployed appropriately, can do a lot of communicative work—from requesting the time to conveying the size of a drink order. You see this most obviously when vocal-tract calisthenics are of no use, like when a person's mouth is full or when he or she doesn't speak the local vernacular. But physical gestures are also deployed as intentional communicative acts of their own. An A-OK gesture tells a pilot he's cleared for takeoff. A Check-Please gesture summons an attentive waiter. And the Bird, well, you know what that does. Across a room, across the world, across the lifespan, people silently convey information using visible movements of their bodies. Words tell only part of the story of how we communicate; gestures tell the rest.

Gestures like those mentioned above are so rich with conventionalized meaning within a culture that they can replace words. This also makes them

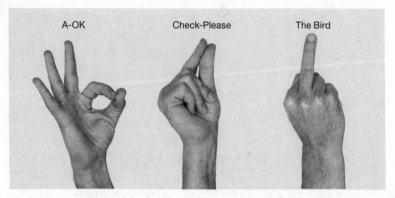

Source: David Bergen.

relatively easy to detect. But these emblematic gestures are merely the tip of the manual iceberg. Most speech is accompanied by often subtler and unnoticed gestures. Sometimes a finger can provide information redundant with the words it accompanies—a contestant on a dating show might punctuate the words *I choose Mary* by pointing to the lucky winner. But movements of the hands, head, and torso can also encode information beyond what's strictly conveyed by words. For instance, suppose someone with a wry sense of humor says, "Oh yeah, I had a great time at the opera." Did she really? Or is she being sarcastic? Her body might tell you. Suppose she accompanies the words with a roll of her eyes and a flick of her wrist in the form of the Jerk-Off gesture on the next page. Probably not an opera lover.

People often gesture when it's useful to the person they're speaking to, like when giving directions. But they also gesture when it's not, like when there's no one to see them. You've probably caught yourself gesticulating when talking on the phone or when staging imaginary conversations in the shower (telling off some self-important PTA member or delivering your Nobel Prize acceptance speech). People gesture even when it couldn't possibly benefit listeners because the listener is a newborn infant or a blind person.[2] Gestures like these that accompany and complement speech will make up the preponderance of the communicative body movement that you, the Martian anthropologist, will notice.

But unlike you, Martian anthropologist, we mere humans only rarely take conscious note of all this vigorous activity of the arms, head, and torso. One way this manifests is that we rarely consider gestures important enough to enshrine in written language, with the exception of certain

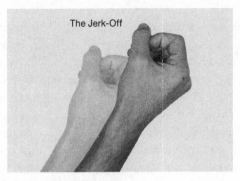
The Jerk-Off

Source: David Bergen.

emoticons like ‾_(ツ)_/‾ (which is supposed to be a shrug, but is really rare because try typing that on your phone!). Very occasionally, you'll come across gestures transcribed in words, like *shrug* or *sigh*, but these are vanishingly rare. A more recent innovation, emojis, can encode limited gestures, like Thumbs-Up, A-OK, and even the Bird. Nevertheless, these remain limited to certain users and contexts. Gestures are mostly absent from written descriptions of pretty much any human interaction. For example, scripts and screenplays contain lots of words for people to say but only the occasional direction regarding gesture. And even when gestures might matter most to people's lives, in court transcripts, they're again mostly absent or at best vague. For instance, consider this example of courtroom dialogue from the Alaska Shorthand Reporters Association:[3]

Q. Did you see the driver of the other car?
A. (Nods head)
Q. Can we have an audible answer, please? The reporter can't take down a nod or shake of the head.
A. Yes.
Q. How tall would you say the other driver was?
A. About this tall.
Q. That's about five-foot-eight?
A. No. More like six feet.

Notice how problematic gestures are here. Court stenographers are the best real-time transcribers of language known to humankind, but even they can't encode everything important and meaningful about gestures. On the rare occasion when a gesture does make its way into the written record, it still remains vague—for instance, "(Nods head)." From a description like this, it's impossible to know if it was a nod with conviction, a hesitating nod, or any other kind. Because the head nod could convey information about the witness's certainty, it could be invaluable to the proceedings. But even in court, the overwhelming majority of gestures go unrecorded. The way a witness shrugs her shoulders or scrunches her eyes, the trajectory she uses with her hand to depict how a car came to an abrupt or careening stop— gestures like these mostly don't make their way into writing because they rarely permeate our consciousness. In short, we largely treat communication as primarily about words, with gestures being optional add-ons.*

Those gestures that we do notice tend to be the profane ones. For example, a lake of ink was spilled during Barack Obama's first presidential campaign as political observers asked, did Obama just flip someone the Bird? On April 17, 2008, the *Los Angeles Times* observed that in a speech, he scratched his face with his middle finger while describing Hillary Clinton's debate performance.[4] And it happened again during his victory speech in November of the same year while he was praising his defeated opponent, John McCain.[5] We can't know whether his middle finger betrayed what he really thought about his political opponents or whether his nose just itched. But his finger had a way of riling people up.

To understand how humans communicate, we have to tackle gesture. And many of the same things one might want to know about words are also important to ask about gestures. What do they mean? Where do they come from? Why do we use the gestures we do? How similar and how different are they across cultures?

Taking a cross-linguistic, cross-cultural view—the same strategy we adopted when asking questions about words—most clearly reveals the answers to these questions. The trick is to find gestures that do roughly equivalent work in each language—that have largely homologous meanings. When we applied this strategy to words, we noted that the word *fuck* translates into foreign words that are as different as they can be; nothing about the sound or spelling of French *baiser* or Chinese *cào* makes them better or worse words for that particular meaning than any other sequence of sounds or letters.

---

* This could be a positive feedback loop. We might fail to write down gestures because we don't think they're important, and we might think they're not important in part because we don't have easy ways to write them down. I do hope someone will figure out what causes what.

But with gesture, finding these equivalents is more challenging. Take just the earlier examples. Many cultures don't have a specific gesture for calling a waiter—because this act is so dependent on a particular type of social interaction. Same with the A-OK gesture. And the list goes on. There are few equivalents around the world for familiar North American gestures like the Loser (an *L* on the forehead) or the Chicken (bent elbows moving up and down to depict chicken wings, among various other manifestations).[6] Likewise, it's easy to find examples of gestures native to other cultures that would be unfamiliar in North America. For instance, in France there's a gesture called Quelle Barbe ("What a Beard!"), in which the backs of the fingers rub the side of the cheek (in the beard location). It means something like "boring." The closest American equivalent might be Whoopdeedoo, where an upward-pointing index finger describes a circle in front of the body. Or the best approximation might be Twiddling-Thumbs. But neither is exactly right. Whoopdeedoo generally indicates the unimportance of whatever's under discussion rather than boredom experienced by the gesturer. And my sense is that Twiddling-Thumbs indicates inaction and impatience more than pure boredom.

Another French gesture without a clear local analog is On Se Tire ("Let's get out of here"), which also appears in Italy and elsewhere in southern Europe. You can see it on the next page. There's not really much in North America or, as far as I can tell, in most places around the world to compare this to directly. The closest thing here to On Se Tire might be Round-'Em-Up, which actually looks a lot like Whoopdeedoo—index pointing upward, describing a circle.* But Round-'Em-Up appears to be much less widespread than On Se Tire.

We can already see that the conventional gestures in languages convey diverse meanings. This starts to answer the question about how universal gestures are. In absolute terms, they aren't universal in either form or meaning. This diversity of gestures around the world also makes it hard to answer the second-order question: In those cases where you do find gestures with similar meanings across languages, how similar do they look?

To answer this question, we have to find meanings that gestures are more consistently deployed to encode in the world's languages. Gestures get used for a small set of very common things. One of these is pointing.

---

* It seems to me, from scattered observations, that Round-'Em-Up is generated by rotation at the elbow, whereas Whoopdeedoo comes more from the wrist. But this is only a hunch.

French gesture On Se Tire.
Source: Sylvain LeLarge, www.talenvoortalent
.nl/englishspeakers.pdf.

People point differently in different places; in Japan, you point to yourself by putting your index finger to your nose;[7] in parts of Papua New Guinea, you point with your nose![8] But pointing appears consistently. Another of the usual suspects is using the hands to depict space—to show the size or relative locations of things. People across cultures also gesture to greet and beckon to one another. And finally, around the world people use gestures to offend.

Naturally, we're going to focus on the last of these. And so we ask, how do people around the world use gestures to insult, to demonstrate disdain, to deprecate? What movements of the body are offensive and why? How similar are the profane gestures of the world's languages? And do any universal principles govern them? To answer these questions, we go on a tour of Birds of the world.

$ % !

Let's begin with the basic facts. The Bird (or the Middle Finger) is of course a big deal in North America. It's our most censored and most disputed gesture because it lives at the intersection of high frequency and high offensiveness. The Bird has varied uses, but these largely track with what the expressions *Fuck you* and *Fuck off* can do. Like these, its linguistic analogs, it can be aggressive or dismissive, but it can also be used jocularly.

The association in people's minds between aggression and extending this one particular finger is strong. We know this from experimental work. One study asked people to extend either their middle finger or their index finger while reading a passage.[9] The passage ambiguously described a made-up person, Donald, who could be interpreted as either aggressive or justifiably assertive (for instance, he refuses to pay his rent, but only after his landlord fails to make repairs—aggressive or assertive?). People extending their middle finger rated Donald as significantly more aggressive than did people extending their index finger. So one finger—as long as it's the correct finger—can change how aggressively you interpret people's behavior.

The Bird has spread throughout the world, at least in part through the penetrating cultural influence exerted by American media. And yet, if you take a trip beyond our borders, you'll find that in many places the Bird won't fly. In some regions, the middle finger is just another digit to count or point with. For instance, in East Asia, the middle finger has traditionally had no notable profane association (although in recent years the Bird has been spreading its wings there too).

Instead, around the world, there exist local Birds with different colors and plumage—gestures that convey aggression and disdain differently from the Bird. Some of these endogenous analogs look like close cousins of our Bird. For instance, the British equivalent uses both the middle and the index fingers in a *V*-shape. (Why the Brits need two fingers where we need just one is beyond the scope of our consideration here.)

You can detect a family resemblance to the Bird in the Up-Yours gesture (also known as the Bras d'Honneur, French for "Arm of Honor") used in southern and western Europe, among many other places. In it, the fist of the dominant hand rises, palm inward, often emerging from under the nondominant forearm. This gives it a similar overall shape to the Bird but using different body parts on a larger scale.

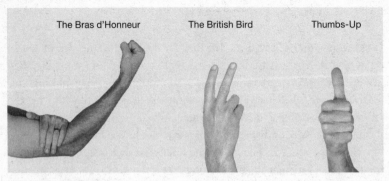

The Bras d'Honneur   The British Bird   Thumbs-Up

Source: David Bergen.

And if you want to stretch the comparison, you might find some simi-larity between these gestures and a profane one used in Iran and Afghani-stan,[10] among other countries, which looks a lot like our Thumbs-Up. Like the Bird, it uses an upward-pointing digit, although instead of the middle finger, it's the thumb. This gesture is usually interpreted as indicating a thumb up somewhere very specific, a place where a thumb could be sur-prising and/or uncomfortable.

And also in the realm of plausible similarity is a Russian gesture (used elsewhere in eastern and southern Europe as well) that looks a lot like what Americans do when we pretend to steal a child's nose. This, the so-called Fig, with the thumb sticking out between the curled index and middle fin-gers, is a slightly milder version of our Bird.

But as we continue our tour, we find gestures that are less and less sim-ilar in overall shape and detailed morphology to the Birds we know—ges-tures that don't extend a finger or fist upward. Brazil has a gesture that uses the handshape of our A-OK (thumb and index forming a circle, with other digits extended) but orients the palm toward the gesturer's own body, with the outside of the thumb-index circle pointing outward. You can see an example on the next page. This gesture is a profane analog of our Bird—it's the rough manual equivalent of *Fuck you*. Or take the Mountza, an offen-sive and denigrating Greek gesture formed with all five fingers extended and the palm exposed. It looks a lot like the Talk-to-the-Hand gesture in North America but has the referential force of the middle finger.

These differences in the ways people around the world use their bod-ies to communicate are important. In practical terms, as a visitor to some foreign country, you generally don't want to accidentally give someone the

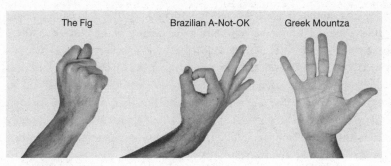

The Fig   Brazilian A-Not-OK   Greek Mountza

Source: David Bergen.

local equivalent of the Bird. Conversely, you do need to know how to manually convey forceful meaning even when you don't speak the local vernacular, whether it's to a cab driver who tries to overcharge you or a maître d' who refuses to seat you. That's when a finger (or two) really proves its worth. But the world's remarkable diversity of profane flicks of the wrist also starts to reveal—as we'll see in a moment—why gestures take the particular forms they do.

$ % !

You'll recall that with respect to words, the different or similar ways a word is translated across languages provide evidence on how arbitrary its sound is. We know, for instance, that the two consonants and one vowel of *fuck* don't have any special relationship to the meaning they combine to convey, and we know this in part because other languages use totally different sounds to convey the same meaning—French *baiser*, Spanish *cojer*, Chinese *cào*, and so on. This is the principle of arbitrariness. Modern English has the word *fuck* because hundreds of unpredictable little things happened over thousands of years to create just the conditions for that word to emerge and be shaped to the point that it means just what it means and sounds just the way it sounds.

When we ask the same question about gestures, we get a slightly different but equally complicated answer. Compare the Bird, the British Bird, the Fig, and the various other ways people use their hands to display disdain and to denigrate. Are these gestures arbitrary, in the same way the words of the spoken languages they accompany are? They're certainly articulated in different ways. The Bird has a totally distinct form from the

Fig, for example. It uses a different handshape and a different palm orientation. (And that's not even considering the variants of the Bird—one where
the middle finger erupts from a closed fist and another where it's flanked
by the bent knuckles of the index and ring fingers.) The British Bird uses
one hand and two extended fingers. The Up-Yours uses two hands and no
extended fingers. The Greek Mountza and Brazilian A-Not-OK are even
more different. It seems that, at least to a first approximation, if diversity
of forms across languages and cultures demonstrates arbitrariness, then
gestures, just like words, are arbitrary.

But let's add a wrinkle to that reasoning. Is it possible that, while diverse, at least some of the Birds of the world are nonarbitrary, each in its
own way? In other words, is there some meaningful reason why the Bird
has the form it does and another reason why the Greek Mountza has the
form it does, different though it might be?

One way to answer this question is to look at the history of each gesture. Perhaps the origin of a gesture reveals why it looks the way it does.
This is easier said than done—gestures don't leave the same paper trail that
words do through written language, and as a result differing stories often
develop about how a gesture came to be. So it can be challenging to dis-
criminate the true history of a gesture—its etymology—from the invented
"folk" etymologies that people propagate. For instance, some version of
this folk etymology of the Bird might have appeared in your inbox:

> In preparation for the Battle of Agincourt in 1415, the French, anticipat-
> ing victory over the English, proposed to cut off the middle fingers of
> every captured English soldier. Without their middle fingers, it would be
> impossible for the English to draw their renowned longbow, rendering
> them incapable of fighting in the future. The English longbow was made
> of the native English yew tree, and the act of drawing the longbow was
> known as "plucking the yew" (or "pluck yew").
>
> But to the great bewilderment of the French, the English were vic-
> torious, and they began mocking the defeated French by waving their
> middle fingers at them, saying, "See, we can still pluck yew!"
>
> Since "pluck yew" is rather difficult to say, the difficult consonant
> cluster at the beginning has gradually changed to a labiodental fricative f,
> and thus the word is often used in conjunction with the one-finger-salute.
>
> It's also because of the pheasant feathers on the arrows used with the
> longbow that the symbolic gesture is known as "giving the bird."

This is a fantastic story—fantastic in that it's total fantasy. Basically nothing about it is true, from the military origin of the gesture to the *pluck yew* contrivance to the timing of its invention to the reason we call it the Bird.[11] To find the true history of the Bird, we'd ideally want to find records of it in visual representations, like paintings, or, failing that, in written descriptions. Perhaps for self-evident reasons, profane gestures are entirely absent from early paintings and drawings. And they tend to be only sparsely accounted for in writing. Fortunately, the Bird is about as notable a gesture as there is, and it has left a discernible trickle of a written record.

Here's what we know from that record. The Bird has had a long and appropriately turbulent flight. It was not invented by English speakers—British or American. And it doesn't date from anywhere close to as recently as the fifteenth century. That estimate is off by about 2,000 years. The earliest records place it in ancient Greece.[12] For example, it shows up in the bawdy Greek playwright Aristophanes's 419 BC play *The Clouds*, in which Strepsiades presents his middle finger to Socrates before waggling his penis at him.[13] Those Greeks could party. In Laertius's *Lives of Eminent Philosophers* (from 330 BC), the philosopher and critic Diogenes expresses disdain for Demosthenes, a prominent Greek statesman and orator, by flipping him the Bird and calling him a demagogue.[14]

So the Bird was around in ancient Greece. The Romans' passionate cultural appropriation of all things Greek extended beyond religion, democracy, and attire into things that really matter, like vulgar gestures. So enamored were they of the Greek Bird that they gave it a name: *digitus impudicus*, the "indecent finger." Then, like now, it was deployed to great effect. The emperor Caligula reportedly denigrated his subjects by making them kiss his middle finger rather than his hand.[15] Cassius, one of these offended subjects, then assassinated him (though there was a lot of assassinating going on at the time, and Caligula doesn't appear to have been the easiest emperor to deal with, so we can't be sure it was the finger that sealed the deal). In another instance of imperial digital intervention, Augustus Caesar allegedly punished an actor who presented the Bird to a heckling audience member by banishing him from Rome.[16]

So we know that the Bird has been around for more than two millennia and that it wasn't always called "the Bird," at least not in its earliest incarnations. That name is a much more recent innovation, dating from the 1960s. As early as the end of the nineteenth century, people used the expression *to give someone the big bird* as a way to describe hissing at another individual, for instance, a performer or public speaker.[17] From there, the term *bird* appears to have migrated from vocalizations to the manual gesture we now associate the word with. Not before 1967 did *flipping the bird* enter the written record. It first shows up in a music magazine article describing the Grateful Dead's onstage antics.[18]

But how did it come to have the form it does—why the extended middle finger? Some say that, at least in ancient times, the Bird was considered a phallic symbol.[19] Strepsiades makes the relation clear with his juxtaposition of presented middle finger and penis. And the belief continues to the present day. For instance, anthropologist Desmond Morris (whom you might know as the celebrated author of *The Naked Ape*) argues, "The middle finger is the penis and the curled fingers on either side are the testicles."[20] This might begin to explain why the Bird has the shape, or shapes, that it does. This explanation leans on the idea of iconicity—the notion that gestures may look like the things they represent. The Bird looks something like an erect penis.

Similar iconic accounts have been offered for all the profane gestures we've seen thus far. The Up-Yours in fact has the same proposed explanation: it's believed to have originated as a phallic symbol too.[21] The Fig has a more complicated history. In early Italian tradition, its name gave it away—it was described not only as *making the Fig* but also as the *far le fiche*, or "cunt gesture." This is pretty damning evidence that people of the time thought of it as representing female genitalia, and the typical interpretation is that the thumb itself represents a clitoris. But by contrast, in current Russian use, the Fig is called *shish*, or "pine cone," a word also used to refer to the glans, or tip of the penis, perhaps represented by the tip of the thumb. If these names are any indication, the Fig's various incarnations over time and space have been iconic for different body parts.

But it's not all phalluses and clitorises. The Greek Mountza—that's the open palm oriented toward the denigrated person—apparently dates back to a Byzantine penal custom of wiping cinders on criminals' faces to defame them as they were paraded through towns (although it may have precursors in a gesture used to cast curses).[22] The gesture derives its name, Mountza ("cinders"), from the ash-wiping practice that the hand evokes. Similarly, cultural interpreters describe the circle formed by the thumb and index finger in the Brazilian A-Not-OK as representing the anus.[23]

These proposed origin stories are all quite similar in one way. They all explain profane gestures from around the world as more or less analog representations of specific things—usually body parts but also denigrating actions. This "iconicity" is akin to sound symbolism, but it lives in the visual rather than the auditory modality. The erect middle finger of the Bird originates in its similarity to the penis it's meant to represent. The touching thumb and index finger of the Brazilian A-Not-OK form a circle to represent the shape of an anus.

But if we're hoping to understand why gestures have the forms they do, we're still missing a step. These origin stories, even assuming they're correct, only go as far as to explain why people might use an extended finger to represent a penis, or a thumb emerging from a clenched fist to represent a clitoris. But these gestures don't mean "clitoris" or "penis." They don't mean "wiping ashes" or "anus." Like their linguistic analogs, they serve predominantly as forceful indications of disdain or denigration. This is the missing step. Why would a manual representation of a phallus (or anus or clitoris) indicate derision and deprecation? Why would it be aggressive to show a manual facsimile of an organ?

Anthropologists have argued, at least for the phallus case, that it's just one of many examples where "the act of male erection or copulation becomes symbolic of male dominance and can be used as a dominance gesture in totally non-sexual situations."[24] If that's true, it's hard to recognize in the modern world. I suspect you'll probably agree that revealing an actual erect penis would probably be out of place in most situations where someone wants to exert dominance. You wouldn't see that happening at the weigh-in before a mixed martial arts fight or in a presidential debate. Moreover, if you were a supervillain, you wouldn't engender fear in the hearts of interlopers by lining the entrance to your lair with erect penis statues. So even if this is why phallic gestures came to have the function they now have, it doesn't seem to relate to the real-world experiences of people in today's developed world. And the phallic representation explanation really falls limp when it comes to anus- and clitoris-based iconicity. So at best, it's a story about why some of these gestures originally came to have the functions they have—and not for why they continue to have them.

We do know that the various Bird analogs we've reviewed find their ostensible origins in things that are themselves taboo: genitalia, sex acts, and so on. So the best explanation relies again on the Holy, Fucking, Shit, Nigger Principle. Perhaps the same selection pressures that make words about these big four topics most suitable to become profane also take handshapes and body movements about the same topics and groom the best candidates into profane gestures.

But even if these historical accounts are correct in their broad strokes—even if there's a nugget of iconicity in the origins of profane gestures—this still doesn't tell us whether profane gestures remain iconic in the minds of modern language users. The proposed resemblances between fingers and penises (and so on) aren't particularly hard to see—the Bird looks plausibly like a penis. But we should check ourselves to make sure we're not just reading in iconicity where we want to see it. Geometrically speaking, there are in fact a lot of things in the world that, like the Bird and like a penis, are longer in one dimension than in the other two. Likewise, many things are circular. And we wouldn't want to fall into the trap of labeling everything so proportioned as phallic or anal, respectively. So how can we tell—in the mind of a contemporary speaker of English, Russian, or Brazilian Portuguese—when a finger is a phallus and when a finger is just a finger? How can we know that we, along with anthropologists, historians, and indeed the people of ancient Rome, aren't just being drawn in by a simplistic explanation, one that we might be biased toward based on what we know about linguistic and other cultural taboos? How do we know we're not seeing what we want to see?

In essence, I raised this same question in the last chapter about four-letter words. Simply observing a pattern in a language doesn't mean that the pattern also has an internal manifestation in the minds of individual people who use that language. We know that raising a middle finger predisposes you to interpret events more aggressively. But does it also activate thoughts about penises?

There's only one way to answer this question, and that's to flip people off and see if that makes them think about penises. An experiment! The first thing to decide is which version of the Bird you want to show people. As I mentioned above, there are at least two major variants, one where the middle finger erupts alone from the fist and another where it's flanked by half-raised index and ring fingers. An Internet image search for "middle finger" reveals that—if online images are representative of real-world proportions—the large majority of Birds are of the former, lone-finger type. There might plausibly be differences in the detailed interpretation of these two variants—perhaps, as Desmond Morris suggests, the curled index and ring fingers represent testicles in the minds of gesture users. And yet, if people today interpret these gestures as iconic, the middle finger ought to represent the shaft of a penis in either case, so going with the more frequent variant seems like a reasonable approach.

The second big decision is how to detect when a person is thinking about penises. Cognitive psychologists have devised a lot of tools to detect whether a word or concept has been activated in someone's mind—everything from measuring how long it takes that person to read a word to whether or not he or she can solve an anagram puzzle with the word's letters jumbled up. One tool that seems particularly well suited for our task is word completion. Suppose you give people a few letters followed by some blanks, like p e n _ _ . The job of your participants is just to fill in the blanks to make an English word. This particular set of letters has several possible correct answers in English. There's *penis* of course, but also *penny, penal, pence,* and *penne.* The question is whether people are more likely to recognize p e n _ _ as the beginning of *penis* when they've just seen the Bird. If so, that would suggest that the Bird activates the concept of a penis or the word *penis* in their minds.

There's a final decision we'd have to make in designing an experiment like this: determining the control condition. In an experiment, you want to know whether something you do to people (say, flipping them the Bird or giving them an experimental drug) has an effect. But that effect has to be measured by comparison to something else. In a pharmaceutical experiment, the control is usually a placebo—a pill, for instance, that's identical to the one that delivers the drug, except that it's missing the experimental compound. What should the control be in a middle finger experiment?

If the control were just nothing—that is, if people completed the anagram task after seeing the Bird in one condition and after seeing nothing in the control condition—then we wouldn't know if increased penis spotting was due to the Bird in particular or to gestures in general. So a first attempt at a reasonable control condition might be to show people a gesture that doesn't have any plausible association with penises. Like maybe the A-OK.

As it happens, I ran this experiment. I recruited two hundred people to perform a word-completion and gesture-memory task. They all saw the p e n _ _ prompt after seeing a still image of a gesture—either A-OK or the Bird. And I counted how many people in each condition responded with *penis* and how many generated another response, like *penny* or *penal.* You can see what I found on the next page. People who saw the Bird were statistically no more likely to answer *penis* than those who first saw A-OK.[*]

Now, you could reasonably object that I didn't see any difference because the technique simply doesn't work. Maybe I'm bad at science in any

---

[*] There's no significant difference between the conditions, as determined by a Fisher's exact test.

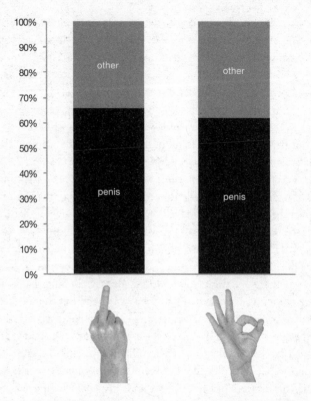

The Bird doesn't lead to significantly more *penis* responses.

of a hundred ways that could have produced a null result. To alleviate this concern, I actually built something into the experiment known as a "manipulation check," intended just to determine whether people's word-blank-filling tendencies could be pushed around using gesture. Here's how it worked. Everyone who answered the p e n _ _ prompt also saw another prompt, p e a _ _, which followed a different pair of gestures. The first was a Peace gesture. And the second was a Thumbs-Down. Overall, slightly less than half of people completed the word p e a _ _ as *peace*. Other popular words to type were *pearl*, *peach*, and *pears*. But critically, as you can see on the next page, people who first saw the Peace gesture were more than twice as likely to type in *peace* as people who saw the Thumbs-Down.* This

---

* A Fisher's exact test reveals a very strong significant effect of gesture on word completion response, $p < 0.00001$.

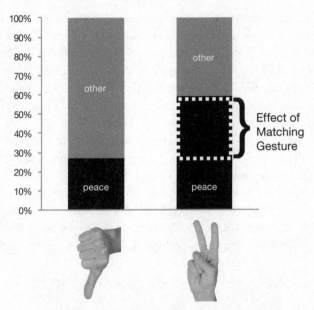

The Peace gesture leads to increased *peace* responses.

successful manipulation check means that the technique works in general. Seeing a gesture can make you think about a word, as measured by how you complete a prompt. So if the Bird indeed makes people think about penises, it ought to have led to more *penis* responses to the prompt. That it didn't suggests that perhaps it doesn't.

Still, you could have other concerns about this result. Here's an alternative account of why the Bird would have no significant effect, which is what we observed. Maybe everyone who participated had exactly the same idea of what p e n _ _ was trying to get at—*penis*. But suppose that in the population, a certain proportion of people simply don't want to type *penis* during an experiment. If this hesitant group comprises 35 percent of the population, that would produce precisely the pattern we saw: two-thirds of people wrote *penis*, regardless of the gesture they previously saw, and one third refused to. How do we know that this isn't what was going on? The answer, as it usually is, is another experiment.

We need a way to determine whether a gesture can get people to think about penises. So why not use a gesture that's definitely about penises, like the Finger-Bang gesture, in which the index finger of one hand moves inside a loop created by the index and thumb of the other hand?

If this doesn't pump up people's *penis* responses to the p e n _ _ prompt, then there's clearly something wrong with the method. Conversely, if Finger-Bang works where the Bird doesn't, that suggests that the Bird simply doesn't lead people to think strongly about penises. So I ran the same experiment as before, but with two changes. First, people saw one of three gestures before p e n _ _ . They could see A-OK or the Bird, as before, or they could see Finger-Bang. And second, I ran the study with more participants—bumping it up to 240—because with participants divided among three rather than two conditions, I wanted to make sure enough people saw each gesture. Two interesting things happened.

First, there was still no significant effect of the Bird. As you can see below if you look at the two leftmost bars, there were slightly more *penis* responses after the Bird, but the difference wasn't statistically reliable. This replicates the finding from the first experiment. Second, and this is the new

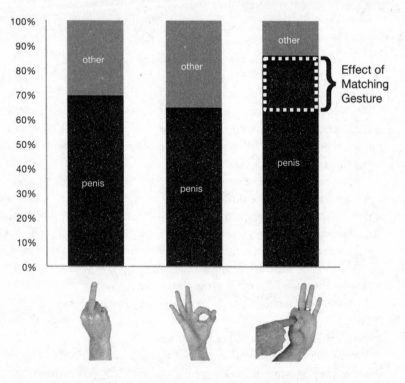

The Bird doesn't lead to increased *penis* responses, but Finger-Bang does.

thing, Finger-Bang did significantly increase *penis* responses, by about 20 percent.*

The interpretation is pretty clear. This technique is sensitive enough to detect when gestures make people think about words or concepts. And although Finger-Bang makes people think about penises, the Bird does not. In its ancient history, the Bird may have originated as an iconic representation of a penis. But that association appears to have died off.

$ % !

So does this mean that gestures like the Bird are arbitrary? Or are they iconic? In a sense they're arbitrary. When you look at the simple mapping between form and meaning, nothing about a middle finger looks like the concept it conveys—roughly, in words, *Fuck you*. And an extended middle finger doesn't convey this notion any better or worse than any of the other variants we see across the world. So, unlike *cock-a-doodle-doo*, there isn't a resounding similarity among the Bird equivalents across the globe. And as we saw from the experiment results, seeing the Bird doesn't appear to lead people to think about the word *penis* or about penises in general.

But at the same time, there's a way in which these gestures are less than entirely arbitrary. Although they don't look like what they mean, they often look like something else on which that meaning is based. Like a penis. Or other things. But usually a penis. Words denoting genitals and bodily functions often come to have profane functions as well (the Holy, Fucking, Shit, Nigger Principle at work). And similarly, gestures that historically derive from imagistic representations of genitals and their functions take on profane uses as well. In a way, that makes them less arbitrary. They look the way they do not by chance but by design. The fingers, the fist, and the palm were selected to represent things that they look like. And gestures that denote those things that they look like are recruited to perform profane functions. Their ultimate use is several degrees removed from where they originated. But it's not random.

Now, most of the profane gestures we've looked at aren't the most transparently iconic signs imaginable. They require a little interpretation, and the fact that they vary across cultures speaks to the importance of

---

* A two-by-three chi-squared test revealed a significant relation (p < 0.01), and the pairwise difference between Finger-Bang and each other condition was significant by Fisher's exact test (both ps < 0.05).

cultural knowledge. They're subject to conventions. Even assuming that the Thumbs-Up and the Bird are equally iconic, they have different conventional meanings across cultures. So the story is a little more complicated than merely asking whether a gesture (or word) is iconic or arbitrary. Even if it's iconic, we also have to know how transparent it is. Some gestures might be so transparently iconic that anyone in the world, even without specific knowledge of the language and culture that they derive from, could still figure out what they mean. Other gestures might require extensive familiarity with cultural conventions that users of that gesture are party to.

And when we dig a little deeper into profane gesturing, it's clear that there exist other gestures that are far more transparently iconic than the Bird, the Fig, and their ilk. Consider, just for the sake of illustration, gestures representing sexual intercourse. We've already seen the Finger-Bang gesture, where one extended finger (often the index or middle finger) of one hand moves in and out of a circle formed by the other hand (usually the thumb and index finger but occasionally the whole fist). Another is a gesture I haven't seen described in print, but let's call it the Fist-Thrust. It uses a fist, usually palm down, pumping away from the body and then back toward it repeatedly. And then there's the Pelvic-Thrust, where both elbows are bent and pump backward past the hips while the pelvis thrusts forward. And of course there are others. Each of these is more transparently iconic than the Bird. There's more detailed shape information about more of the scene. That makes them easier to interpret independently of convention. Moreover, there's more room for individual variation without compromising the message. With the Finger-Bang, the dynamics of the finger entering the circle formed by the other hand can, if the gesturer so desires, convey details about the dynamics of the represented sex act.

I'll leave this line of argumentation here, because I think the point is probably made. Profane gestures like these, gestures at the most transparent end of the spectrum, look far more like what they're meant to denote than the Bird does. And not surprisingly, we have limited experimental evidence, at least for Finger-Bang, that they activate words for the represented genitalia in the minds of language users. You can think of iconic gestures like these as the manual analogs of spoken onomatopoeia. Just as onomatopoeic words of spoken languages imitate sounds, so gestures can imitate actions, as these do.

And this easy activation in the mind of the observer may be a communicative edge that explains why we have vulgar gestures in the first place and why they tend to be iconic. Profane words, as we've seen, generally don't resemble what they mean. Profane gestures, by contrast, often do. This makes them more direct, more evocative triggers for the concepts they convey than words often are.

To sum up what this tour of profane gestures has revealed, we now know that the profane gestures of the world vary and that they find their origins in areas like sex and bodily functions that are also, not coincidentally, the sources of taboo words. Profane gestures are largely more iconic (to different degrees) and more transparently so (in different ways) than typical words of a spoken language, and this can give them a leg up in directly activating what they refer to.

But we've only witnessed the beginning of the hands' power to offend. Consider that there exist entire languages that are articulated, like gesture, via visible movements of the hands, arms, torso, head, and face. These signed languages are the more sophisticated, more articulate siblings of the gestures we're familiar with. Gestures are isolated communicative bits, which limits what they can communicate. You can use a gesture (for instance, many of those discussed in this chapter) to start a fight, but gestures alone won't allow you to talk yourself out of a fight by explaining how your anger-management issues stem from repressed feelings of low self-worth. You can use a gesture at a physics convention to summon someone over, but you can't use gesture alone to make advances on that person by showing off your quantum mechanics chops. Gestures are expressively impoverished compared with the words of fully formed languages.

So we should probably expect signed languages—which harbor all the expressive potential representative of full human languages in the visual modality—to set the standard for manual obscenity. But in addition, signed languages hold the key to why gestures are so much more iconic than words are. Gestures differ from words in several ways. The first is the modality: words create a predominantly auditory signal, but gestures are mostly visual. Maybe vision and movements of the body are better suited for iconicity than sound is. But there's another possible factor at play. Words are different from gestures because they're an integrated part of a communication system that allows them to be combined to express any thought. Signed languages can uniquely tell us whether the increased iconicity of profane (and other) gestures has to do with their visual nature or whether words are just more arbitrary as a consequence of being the building blocks of a systematic language.

$ % !

Millions of people around the world communicate primarily using a signed language, of which there are hundreds. Most signers are deaf or hearing impaired, but some hearing people—usually relatives, friends, or associates of deaf people—also sign. Signed languages share one big thing with gestures: both deploy visible movements of the hands, arms, torso, and face. But in most other ways—their structure, their complexity, their expressiveness—there's no comparison. Signed languages are fully functioning languages. Like the spoken languages you're likely more familiar with, they place strict constraints on how to articulate words,[25] and they have inviolable, meaningful, and abstract rules of grammar.[26] Let me give you some examples.

Compare the two signs from American Sign Language (ASL) on the next page, BITCH and BASTARD. I might need to explain why the names of signs are set in all caps. Signed languages aren't just signed versions of local spoken languages. So it would be inaccurate and often misleading to label signs with their translations into some spoken language (like English). But still, we need some label for the signs so that we can talk and write about them. The compromise is to label them in all caps typically using words of the local spoken language, when there's a close translation equivalent. The ASL sign that we label PUSSY below means something similar to English *pussy*. But you'll see examples where the sign labels aren't recognizable in English. With that out of the way, let's look at BITCH and BASTARD. As you can see, both use the same handshape—a flat palm— and both involve striking the face, but they do so in different places.*

Everything about how you form these signs with your hand and arm is strictly regimented to correctly articulate them. If you bend your fingers just a little, or if you touch your palm to a different part of your face, or if you touch it to your face and hold it there rather than tapping briefly, you will be "mispronouncing" the sign. You may inadvertently sign another

---

* And a presentational note: Signs in a signed language involve hands configured into particular shapes moving through space and changing shape over time. This means that the best way to show a sign is in person or, barring that, using video. But this is a book, and writing is ill-suited to describing how signs are articulated. I'm just as disappointed as you are that I can't embed video in this book. It's 2016. Come on. So instead, I've done the next best thing. The signs you see in this chapter encode motion using sequences of still images that you should read like a comic strip.

BITCH in ASL.
Source: Jolanta Lapiak.

BASTARD in ASL.
Source: Jolanta Lapiak.

word—for instance, the only difference between BITCH and BASTARD is where you hit the face—or you may produce gobbledegook, in the same way that a small change to the pronunciation of a word in a spoken language will change meaningful *bitch* to meaningless *gitch*.*

Signed languages also have very specific conventional rules of grammar that dictate how the signs fit together to form larger utterances. American Sign Language has its own grammar, totally distinct from that of English. Let's compare an English sentence with its equivalent in American Sign Language. Say you want to tell someone she's an unlikeable person. In English, the sentence might follow the typical subject-verb-object order of transitive sentences: *You are a bitch*. But in American Sign Language, as you can see below, the sentence would more probably go like this: YOU BITCH YOU. There's no verb. But the subject occurs twice, at the beginning and the end. This most definitely isn't English, but it's still grammar—standard grammar for ASL.

The point here is just that ASL, like any signed language, is a fully formed language with its own rules, distinct from the spoken languages around it. And this gives it an expressive power that far surpasses speech-accompanying gestures like the Bird. And signers swear. There's been surprisingly little research on the profanity of American Sign Language—or any other signed language for that matter. But here's what little we do know about swearing in sign, largely taken from the primary resource on the topic, a paper by Gene Mirus of Gallaudet University and colleagues.[27]

YOU BITCH YOU in ASL.
Source: Jolanta Lapiak.

---

* In the time since I generated this example, I've been informed that in some varieties of mostly Canadian English—perhaps centered in Saskatchewan—*gitch* is in fact a word, meaning "undergarment." This strikes me as a suitable meaning for a closed, monosyllabic word like *gitch*.

Physical characteristics are fair game. Many English speakers are shocked when they first learn how casually Mexican Spanish speakers describe people by their physical characteristics. If you have a high body-mass index, you might well be nicknamed *gordo* ("fatty"). If you have exceptionally large ears, people might call you *antenas* ("antennas"). The same is true in American Sign Language. As Mirus and colleagues put it, "An ASL signer might pick someone out by their large nose, acned skin, or asymmetrically placed eyes. . . . This is perfectly acceptable behavior; it is not rude or even politically incorrect, regardless of the situation."[28] So taboo language in ASL doesn't typically derive from these sources, unlike in English.

Audiological status can be inflammatory. Many of our most profane expressions are terms that describe groups of people. One group that ASL signers find socially important enough to have slurs for is hearing people. For example, there's a sign in ASL for HEARING in which the index finger makes circles in front of the lips, perhaps to indicate that hearing people communicate by moving their lips. There's also an insult built off this sign, in which you take the index finger and move it up to the forehead to signify THINK-LIKE-A-HEARING-PERSON. According to Mirus and colleagues, this sign is derogatory and degrading.[29]

How you sign it makes a sign taboo. Many profane signs can also be used in nonprofane ways. For example, the sign PUSSY (which we'll discuss more in a moment) looks almost identical to the sign VAGINA—same hand shape, same location. They're distinguished only in that the former is produced with "a quick, sharp movement and sometimes an angry (or perhaps joking, depending on the situation) facial expression."[30]

Finally, signs are pretty iconic, but it's complicated. Gestures tend to be more iconic than the words of spoken languages and more transparently so: they're more likely to look the way they look because of what they mean. So are signs. Even if all you know about American Sign Language is BITCH, BASTARD, and YOU, you might already have a pretty good sense of how arbitrary its signs are. Some, like YOU, which you've just seen, are not at all arbitrary. Many other ASL signs are similarly not only iconic but transparently so.[31] This is especially true of profane signs. On the next page you'll see the signs FUCK and PUSSY in ASL. I've left them unlabeled to let you experience for yourself precisely how iconic they are or aren't.

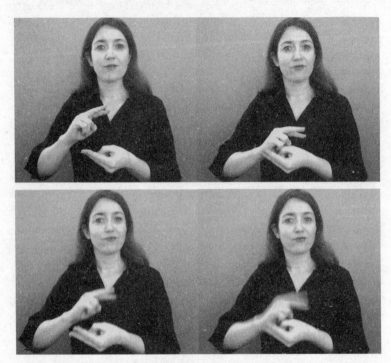

A sign in ASL. Source: Jolanta Lapiak.

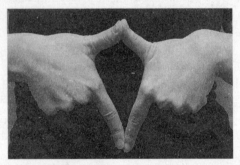

Another sign in ASL. Source: Jolanta
Lapiak.

I presume you had no trouble ascertaining which is PUSSY and which is FUCK. The iconicity of PUSSY would be hard to miss. Remember, for this sign to be PUSSY and not VAGINA, it needs to be accompanied by the right movement and facial expression. In contrast to PUSSY, you might have trouble seeing why FUCK looks the way it does. It may help to know that in this sign, as elsewhere in ASL, the outstretched index and middle finger represent legs.

While PUSSY is quite transparently iconic, other signs, like BITCH and BASTARD, are less obvious. What BITCH denotes—an aggressive or unpleasant person—doesn't superficially have anything to do with touching the palm of the hand to the chin. Same with BASTARD and the forehead. And yet, the forms of these signs aren't entirely arbitrary. If you happen to know a lot of American Sign Language, you might have noticed that BITCH and BASTARD actually make sense in terms of how the rest of the language uses space systematically. Signs in ASL for females, like GIRL, MOTHER, AUNT, and so on, tend to involve touching the chin. And signs for males, like BOY, FATHER, UNCLE, and the like, tend to involve touching the forehead.

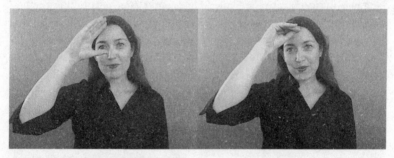

BOY in ASL. Source: Jolanta Lapiak.

GIRL in ASL. Source: Jolanta Lapiak.

Why? The historical explanation is iconic. The sign BOY originates in touching the brim of a cap, at the forehead. And GIRL comes from the placement of a bonnet string under the chin. Generalization from this pattern may have introduced a local systematicity into the signs of ASL. Just as *gl*-words in English tend to have meanings related to light or vision, and just as profane English words are more likely to be closed monosyllables, ASL has its own systematicities, based on its own conventions and its own history of iconicity. BITCH and BASTARD are consistent with the rest of the system in where they're placed: forehead for male, chin for female. Of course, that doesn't make them any less arbitrary for someone who doesn't already know the language—the sign BITCH is no more or less inherently appropriate for its meaning than is the English word *bitch*. But iconicity and convention underlie even arbitrary-seeming signs.[32]

$ % !

I started by selecting some signs from ASL because it's the largest signed language indigenous to the United States and Canada, as well as the best documented. The number of people who currently use ASL is unknown; reasonable estimates range from about 100,000 to about 500,000 signers.[33] But ASL is just one of the hundreds of signed languages around the world: French Sign Language, Mexican Sign Language, British Sign Language (BSL), and Japanese Sign Language are just a few of the larger and better-studied ones.[34] And most of these languages are unrelated. British Sign Language, for instance, developed along a totally separate track from American Sign Language (which itself derived from a nineteenth-century form of French Sign Language).[35]

But because of their rampant iconicity, even totally unrelated signed languages show easily noted similarities. The sign PUSSY in British Sign Language—which, to reiterate, is totally unrelated to ASL—is identical to the sign PUSSY in American Sign Language. They're both iconic and in the very same way. Slightly less similar are the ASL and BSL signs for FUCK. Compare the ASL sign we saw earlier with its BSL analog.

The two FUCKs exhibit clear differences. They use different handshapes: ASL uses closed fists with index and middle fingers extended, whereas the hands in BSL use open palms with a gap between the thumbs and the other fingers. The motion is also different. It's hard to depict this

PUSSY in BSL. Source: Commanding Hands.

in still images, but whereas FUCK in ASL uses a together-apart-together motion, the BSL version taps the hands together only once.

Are these two signs for FUCK still iconic, even if different in hand-shape and motion? Arguably, yes. Iconicity in signs can be just as nuanced as in gestures. FUCK in both BSL and ASL could be iconic by encoding something about the meaning of the sign in how each is articulated. The meaning of FUCK offers a bounty of details that the form of the sign could highlight and lots of ways to depict those details. The index and middle fingers can stand for legs, as in ASL. Or the gap between the thumb and index finger can represent a crotch, as it appears to do in BSL. Languages have latitude.

So just like in gesture, there appears to be rampant iconicity in signed languages. But as we've seen, this doesn't mean that the world's signed languages are identical. Languages choose how to encode meanings iconically. And even if those choices are all equally valid, once they're made, they're binding—people who learn the language come to take the shape and motion of those signs as given. And it couldn't be otherwise; without settled,

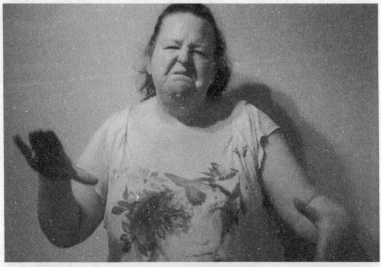

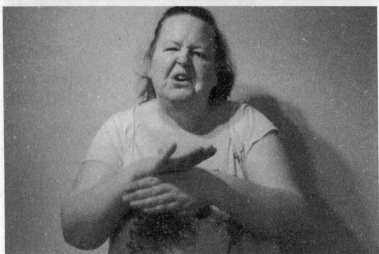

FUCK in BSL. Source: Commanding Hands.

agreed-upon conventions, communication would be reduced to a game of charades. And to be sure, signers are doing something far more complex. Just like speakers of any spoken language, fluent signers can communicate efficiently about anything from the tax code to nanotubes. They can exhort, they can impel, they can request, and they can swear. And they do so—just as quickly as speakers of spoken languages[36]—because the signs of a signed language are just as fixed as the words of a spoken language. Signers use rules of grammar, some of them specific to profanity, just like speakers of spoken languages.[37]

This brief introduction to profanity in the signing world should have revealed two things. First, there are striking similarities across totally unrelated signed languages due to iconicity. The contrast with spoken languages is stark. If you take two unrelated spoken languages—or distantly related languages—the words for similar concepts are likely to be quite different. As we already saw, English *cunt* doesn't sound anything like Cantonese *hai* or Russian *pizdá*. Profane signs are far less arbitrary. And second, while the hundreds of signed languages across the globe are similar in their rampant iconicity, they differ markedly in exactly how individual signs look. This means that while a particular sign in one signed language might tempt the nonsigner to believe that signing is essentially pantomime, it's anything but. Signed languages are conventional systems.

This means that knowing one signed language won't allow you to understand the next one. At its core, BSL is quite different from ASL—so much so that they're usually described as "mutually unintelligible"[38]—just like English and Chinese. When you compare signed languages, everything from the alphabets to the signs and the grammar can be different. For instance, Japanese Sign Language has a sign that's produced by doing what looks like pointing two Birds at the person across from you and pumping them up and down in alternating thrusts. This sign—in Japanese Sign Language—isn't the least bit profane. It means "brothers." In fact, the raised middle finger in Japanese Sign Language doesn't have any sort of taboo connotation. Its meaning might have a hint of iconicity, but the erect finger doesn't represent a phallus so much as a person. In Japanese Sign Language, many signs for people involve extended fingers—MAN is the thumb, WOMAN is the pinkie, and so on. Critically, the extended middle finger is BROTHER. And, again through iconicity, multiple fingers represent multiple people. If you put these principles together, two extended middle fingers represent brothers in a way that's both iconic and conventionalized.*

At the same time, like gesture systems, signed languages balance arbitrariness with iconicity. To say that two signed languages are totally different—because they have different histories, different signs, and so on—is

---

* Thanks to Nozomi Tomita and So-One Hwang for bringing this example to my attention!

only true to a point. In fact, in some ways signed languages are more similar than spoken languages can be to one another. Some native signers of ASL have reported to me that if you are a sufficiently clever and observant signer of ASL, you might do better with BSL than someone totally naive to signed languages. Some signs are similar due to iconicity. Some look superficially different but are motivated in a similar way by iconicity. Perhaps knowing how a signed language harnesses and uses iconicity helps you figure out what's going on in a different language you don't know at all, as long as it operates according to similar underlying principles. And iconicity may also help adult nonsigners learn signed languages.[39]

$ % !

The vocal tract is only one of the channels that humans use to communicate. Most of what you can do with your mouth you can also do with your hands, and vice versa. But the different channels are not equivalent in fundamental and consequential ways. Using your hands and arms and the rest of your body to perform visible actions, whether gesturing or signing, affords different possibilities and imposes different constraints.

Iconicity is one way the channels differ. It's hard to talk iconically about motion and shape using spoken words, but gesture and sign are particularly well equipped to do this because you can use your body to mark out movement through space and trace or recreate shapes. Hands can be contorted into various shapes; they can move in three dimensions with specific speed, dynamics, direction, and so on. Visible movements of the hands afford analog representations of far more of the world than words do. This appears to be the reason that both gestures and signs are generally less arbitrary than the words of spoken languages.

The verbal and manual channels differ in other ways that would lead someone using a spoken language to prefer one or the other in specific contexts. Obviously, there are conditions that make words inaudible and gestures unseeable. On the highway, for instance, the Bird might be the only way to get your message across. Conversely, sometimes undetectability increases the value of a word or gesture. A Bird can be crafted discretely, for example, at the back of a classroom behind a laptop in such a way that the intended audience of other students can see it but the hapless teacher at the front of the room cannot. Gestures and spoken words are processed via overlapping but somewhat distinct pathways in the brain,[40] and it's possible that these pathways give profane gestures more direct access to emotional reactions. And finally, because, as I mentioned earlier, gestures are largely not "on the record"—we take them less seriously as communicative acts than words—they allow plausible deniability. A discrete middle finger scratch of the nose, the type that President Barack Obama has perfected, is ambiguous enough to leave the audience wondering, did he just mean to do that?

4

The Holy Priest with the Vulgar Tongue

Jacques Lordat is the most important neuroscientist you've probably never heard of. Born in 1773, he trained as a physician and then practiced medicine and served as a professor in Montpellier, France. But at the age of fifty-two, he suffered a devastating stroke. A stroke, as you might know, occurs when a blocked or leaky blood vessel reduces blood flow to part of the brain. Deprived of the oxygen that the blood carries, neurons in the affected region start to die off, and this can cause long-term impairments to the functions that rely on those cells. Lordat's stroke apparently stemmed from a tonsil abscess that led to a blockage in the carotid artery,[1] which brings blood to the front of the brain. And it left him unable to speak. But slowly and with effort, he began to heal himself (as physicians will). And throughout his recovery, even in the early days when his capacity for speech was decimated, he could still think lucidly. So when he eventually regained the ability to speak and write, he recorded for the scientific record not only the objective facts of his case but also the subjective experience of what it feels like to lose language due to brain insult. And he eventually used his expertise to lay out the first modern theory of how our brains produce language, a theory that's still broadly recognizable in the contemporary scientific consensus nearly two centuries later.[2] So, in short, he's kind of a big deal.

After his stroke, Lordat focused his research and practice on people who, like him, had suffered language-compromising brain damage. One of his most surprising findings came from a case study, the account of which he published in 1843. In it, he wrote about a parish priest, a man of God, who, like Lordat himself, had suffered a serious stroke.[3] The priest also showed obvious signs of language impairment. Just like Lordat, he retained little ability to speak—only more so. The priest's vocabulary was reduced to just two words. The first was *je*, the French word for "I." Perhaps unsurprising, if you're only going to retain one word. But the other word was, at the time, unimaginable for a parish priest. Lordat wrote that the priest made ample use of "the most forceful oath of the tongue, which begins with an 'f' and which our Dictionaries have never dared to print."[4] That was the French word *foutre*, which by coincidence happens to start with the same letter as its English translation: *fuck*.

In his own experience and that of the priest, Lordat had discovered aphasia—language impairment caused by damage to specific parts of the brain. Unfortunately, aphasia is common. Currently, about 1 million Americans suffer from some type of aphasia due to brain damage at the hands of stroke, as well as traumatic brain injuries, infections, and dementia.[5] And with two centuries of accumulated observations, we now know that many of these people exhibit the very same syndrome that Lordat observed in his priest. They find it difficult or impossible to intentionally articulate and string together words. But some spontaneous language is still preserved— interjections like *yeah* and *huh*, filler words like *um* and *well*, and some of the most vulgar verbal ejaculations in the language.

How can this be? Why does certain brain damage render its victims unable to articulate even the simplest sentences but leave knee-jerk profanity intact? What does this imply about the brain and how it produces language—whether in sickness or in health?

Aphasia is a cornerstone of neuroscience. It's the most revealing tool we have to study the brain's mechanics for generating and understanding language—how something breaks down often provides the clearest window into how it works. And while Lordat's observations and his theory have been influential in the field that he founded, the theories that have built up over the centuries, based in large part on aphasia research, have largely omitted the profanity. With few exceptions, it's become an anecdotal side note.[6] And as I'll argue, that has led to exactly the wrong conclusion about how the brain works.

$ % !

The brain is a roughly three-pound sack of tissue, populated by billions of neurons soaking in a finely tuned chemical bath. The ancient Greeks thought that the brain's primary function was to cool the body—that it served as a sort of organic radiator.[7] In truth, the brain does consume about 20 percent of your body's energy (despite making up only 2 percent of your body weight).[8] But we now know of course that it does more than just produce heat. It also generates light, at least metaphorically speaking, in the form of thought. More relevant for our present purposes, it also allows you to learn, produce, and understand language. When things go right with language, it's because the brain is working. When language breaks down, it's usually due to brain malfunctions. For example, when a well-mannered parish priest spontaneously starts cursing a blue streak, you can bet it has something to do with how language is fleshed out in his brain.

The evidence we can glean from aphasia—from Lordat onward—shows that language is implemented in the brain in a highly structured way. Different parts of your brain are involved in different sorts of operations. Known as localization of function, this is one of the most important principles of brain organization.

Here's how it's supposed to work for language. Suppose you hear someone utter a word.[9] Once the sound hits your ears, the inner ear converts it into electrical signals, pulling apart the sound's different frequencies like the equalizer on a stereo. This electrical signal bounces through a chain of specialized areas in your brain until it reaches the auditory cortex in the temporal lobe, which processes sounds in general. Different parts of the temporal lobe extract information about the speech sounds that make up the word and then send modified signals to a region called Wernicke's area, which is believed to associate the sequences of sounds that you've heard with their meanings. It's thought to be like a mental lexicon. (If the word you've just heard is embedded in a sentence, then there's more work to do, involving other brain areas, but let's leave that out for now.) That's how comprehension is supposed to work. Conversely, to produce language, you might start with Wernicke's area, which could allow you to select words that will adequately express whatever meaning you want to convey. And then signals get transmitted up to Broca's area in the front of the brain, where they're translated into sounds to be articulated.

The evidence that these different areas serve these particular functions—that functions are "localized" in this way—comes originally and most compellingly from aphasia. Damage to the different parts of the

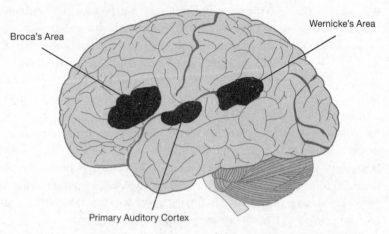

A brain (schematically, with skull removed). This is the outside of the left hemisphere of the brain—the front of the brain is on the left-hand side of the page, and the top is up. Modified from an image licensed under the Creative Commons Attribution—ShareAlike 3.0 Unported License. Source: Hugh Guiney.

brain produces distinct aphasia. When Wernicke's area is damaged, people have trouble understanding language. Moreover, when they talk, they pronounce the words and sentences fine; they just don't make sense. For instance, someone with Wernicke's aphasia might say something like "I did the thing in the thing." It's a perfectly grammatical sentence, but what does it mean? Maybe the speaker wants to tell us she washed the dishes. Or maybe that she seduced an orderly. Who knows?

A Broca's aphasic, on the other hand, uses the right words, and you can usually figure out what he means; it's just that the structure of his words or sentences (or both) is off. He will also have trouble pronouncing speech sounds and the words they make up. So a typical utterance might be "I-I-I make . . . um . . . damn!" With effort, you can sometimes figure out what he's getting at, but the articulation of words is labored. Because Broca's area is often damaged in people who show the symptomatology of Broca's aphasia, it follows that this region subserves the capacities lost in Broca's aphasia. Both Lordat and his priest suffered from devastating damage to this part of their brains—to the point where almost all production of speech was obliterated.

This is a remarkably clean story. If it's right, specific brain regions perform different functions for language—Wernicke's computes meaning, and

Broca's deals with sound. This is the prevailing view of how the brain processes language that you'll see in most introductory cognitive psychology or cognitive neuroscience textbooks (like the one in the footnote at the end of this sentence).[10] And the idea that different parts of the brain do different things (localization of function) is appealing. It's as if the brain houses an efficient assembly line for language.

The problem is that the case of the cursing priest doesn't square with this story. When people suffer from Broca's aphasia, the condition doesn't affect all words equally. It plays favorites. Broca's aphasics often struggle to articulate or remember how to produce run-of-the-mill nouns and verbs. But fixed expressions like *How do you do* or *I can't*, and especially expletives, tend to be preserved. So it's not uncommon to find the language of Broca's aphasics, like Lordat's priest, liberally sprinkled with *Jesus Christs*, *shits*, and *motherfuckers*.[11] Even when patients suffer from nearly complete loss of language—so-called global aphasia—they often retain interjections and profanity. For instance, a recently documented patient could only produce six words: *well*, *yeah*, *yes*, *no*, *goddamnit*, and *shit*.[12]

This same pattern, which shows up in patient after patient, requires a fundamental rethinking of the brain's functional organization for language. It can't be that all language is produced on the same assembly line. Just as neuroscientists have concluded that meaning and articulation are localized separately in Broca's and Wernicke's areas because each can be selectively impaired by damage to the respective regions, so differential impairment of profanity versus the rest of language implies that these too must have different neural bases. The question then becomes, if the parts of the brain that we think perform language functions (like Broca's and Wernicke's) aren't responsible for a frustrated *fuck*, *goddamnit*, and *shit*, then what is?

We uncover clues when we dig a little deeper into what aphasics can and can't say. It turns out that the patient mentioned above who could only say six words (and possibly the priest as well, although Lordat's writings don't record this) could only say his six words reflexively. That is, he could only produce *shit* as an unintentional reaction—a fleeting expletive—but he found it impossible to read or repeat the word *shit* when asked to do so intentionally.

So while it would be noteworthy if we were to discover that profanity is simply generated by different parts of the brain than nonprofanity, there's actually something far more revealing going on. The issue isn't about which words are produced so much as how they are used. It seems that while Broca's aphasia and global aphasia impair intentional speech, they don't interfere with speech that is reactive, impulsive, and spontaneous— so-called automatic speech.[13] This distinction is crucial for understanding how the brain produces language. Automatic speech dissociates from intentional speech—they can be differentially impaired. This means that they must originate in different mechanisms of the brain. A word, when used as an automatic expletive, is generated by one set of brain circuitry; but when crafted intentionally, that same word comes off a different assembly line. Accepting for the time being that Broca's and Wernicke's areas bear much of the burden for intentional speech, we're left to ask, where in the brain do the fleeting expletives of automatic speech come from?

$ % !

As you likely know, the human brain is split down the middle into two rounded, roughly symmetrical halves—its two hemispheres. These two hemispheres do slightly different things. You may have heard of people who are "right-brained" being more artistic and people who are more "left-brained" being more logical. That's mostly hooey, but like a lot of hooey, it's based on a grain of truth.[14] Namely, while everyone with two normal hemispheres uses both of them in parallel to do pretty much anything, from crossword puzzles to rhythmic gymnastics, it is true that your brain does some things in a partly lateralized fashion. That is, some functions use one half of the brain more than the other. And some aspects of language are like this.

In most people, the traditional language centers—Wernicke's area, Broca's area, and so on—are all clustered in one half of the brain. In almost all right-handed people (95 percent), the left hemisphere is more active during language tasks than the right hemisphere. Being left-handed makes it more likely that the reverse will be true—that the right hemisphere will outwork the left, but this is still only true in 27 percent of left-handers.[15]

It might stand to reason, then, that automatic language, like the rest of language, would tend to be produced using the left hemisphere more than the right hemisphere, and more so in righties than in lefties. And the dissociation logic would imply that because automatic speech can be preserved even when Broca's and Wernicke's areas are damaged, automatic language might be generated by some other part of the left hemisphere, outside of the well-known language centers.

A fascinating little bit of evidence on this comes again from people with aphasia. It involves looking not directly at their brains or what words they're able to say but rather at how they use their mouths to articulate words. For this to make sense, you have to know something slightly counterintuitive about how the brain controls physical actions. The left hemisphere of the brain is primarily responsible for sending signals to muscles on the right side of the body—the "contralateral" side—and the right hemisphere of the brain controls actions on the left side. This contralateral control includes moving your hands and feet, and it also includes moving your mouth; the left hemisphere takes the lead in controlling the right side of your mouth, and the right hemisphere is in charge of the left side. Because most people's brains are predominantly left-lateralized for language, they (and that probably includes you) actually speak using the right side of the mouth slightly more than the left side.[16] So when you open your mouth to say a word, the right side tends to open slightly wider than the left side. And it's not just that you're leaving the right side of your mouth slightly more agape overall—the total distance moved on the right side is farther than on the left side.

But that's not true of everything you do with your mouth. On the next page you see what happens when you measure mouth lateralization as people (in this case, people with Broca's aphasia and damage to the left hemisphere of the brain) perform a variety of language tasks. Some are purely linguistic, like generating lists of words or repeating words; others are less so, like singing and smiling. The researchers measured the relative aperture of the left and right sides of the patients' mouths as they moved them in all these different ways. And what you see is striking. These people opened the right side of their mouths more for most language tasks— generating lists of words, repeating words and sentences, saying whatever came to mind, and so on. This indicates more left-hemisphere control. But automatic speech (labeled here as "serial speech")—things like counting from one to ten or reciting the alphabet—induced larger apertures on the left side of the mouth, suggesting greater right-hemisphere control. Like swearing, counting and recitation are often preserved in Broca's aphasia, which is why they're considered "automatic speech."

The relative opening of the left versus the right side of the mouth indicates which hemisphere of the brain is doing more work in controlling speech, and these results show that while most speech is left-lateralized in the brain (because the right side of the mouth is open more), automatic

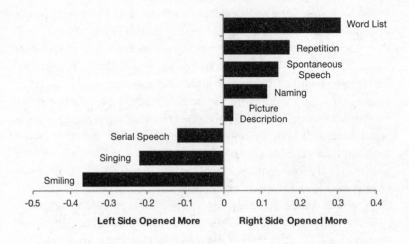

Relative frequency of larger openings on left and right side of the mouth in twenty Broca's aphasics with left-hemisphere impairment. The right side is opened wider during all language tasks except singing and automatic language. Produced on the basis of data in R. Graves and T. Landis (1985).

speech is right-lateralized. Now, the researchers didn't elicit spontaneous expletives. But if these pattern with other automatic speech, they might actually have a right-hemisphere origin.

How can we pursue this idea? The best type of evidence for the lateralization of spontaneous profanity to one hemisphere or the other would come from people who can only use one brain hemisphere. If those people can use one type of language—automatic or intentional—but not the other, this would suggest that the nonworking hemisphere is necessary for the impaired type of language. And hard to believe though it might be, there are in fact people who can use only one hemisphere of their brains—because one hemisphere is all they have. These are patients who have had one hemisphere partially or completely removed for medical reasons.[17]

One such individual was a patient pseudonymized as "E.C.," a forty-seven-year-old, right-handed man who developed severe symptoms from a substantial tumor in the left hemisphere of his brain. As you might expect, since each hemisphere of the brain is the primary driver of motor actions on the opposing side of the body, these included not only aphasia but also motor deficits specifically affecting his right hand. He was admitted to the hospital in March 1965, and the tumor was removed from the left hemisphere of his brain. But the tumor turned out to be malignant, and as E.C.'s symptoms did not improve after surgery, the decision was made in December of the same year to remove the entire left hemisphere of his brain. Given how radical this procedure is—removal of an entire half of the brain—his prospects must have been bleak.

So here we have someone who—if language is left-lateralized—should have no ability to speak. And indeed, after surgery, most of his language capacities had been obliterated. But not all. A report of his condition reads as follows: "E.C.'s attempts to reply to questions immediately after operation were totally unsuccessful. He would open his mouth and utter isolated words, and after apparently struggling to organize words for meaningful speech, recognized his inability and would utter expletives or short emotional phrases (e.g., 'Goddamit!'). Expletives and curses were well articulated and clearly understandable."[8] Let me just highlight the important bits. He was totally unable to produce meaningful speech. Then he got frustrated and swore: "Expletives and curses were well articulated and clearly understandable." Just as a reminder, this person was missing the entire left side of his brain. This is an astounding discovery. You don't need your left hemisphere to talk, as long as you're swearing in frustration. It seems that part of the reason that Broca's and even global aphasics still swear is that circuitry located in the right hemisphere can generate automatic speech.

What is that circuitry like? Are there right-hemisphere homologues of Broca's and Wernicke's areas, specialized for automatic speech? Is it some different type of machinery entirely?

As we narrow the scope of the question from which hemisphere is involved to which regions within that hemisphere, the type of evidence we need also becomes more specific. The critical evidence would have to come in the form of people who have suffered damage to different parts of the right hemisphere of their brain. You look to see whether any of them has trouble only with automatic speech but not with the rest of language—basically, the opposite of Lordat's priest. And you use their language dysfunction to home in on the regions in the brain that house the circuits that generate automatic speech. So the challenge becomes finding people who, due to localized right-hemisphere brain damage, lose the ability to spontaneously produce expletives.

In the literature on aphasias, there are no such people. At least not "people" in the plural. There's just a person. One of them. I present for your consideration that one case. A 1993 report in the journal *Neurology* describes a patient who—like many—had damage to the right hemisphere of his brain.[19] But the behavioral consequences of this damage manifested in a peculiar way. This patient, who was bilingual in French and Hebrew, spoke like a typical speaker of both of his languages. That is, until it came to automatic speech. Reportedly, after suffering brain damage, he was unable to sing familiar songs or recite nursery rhymes, and he couldn't spontaneously swear. And this was most surprising because before suffering brain damage, he was reportedly an enthusiastic purveyor of profanity. Here he is: the mirror image of the parish priest! Instead of preserving automatic language, such as expletives, his brain damage specifically impaired that function alone.

And now for the payoff: Where was his right hemisphere damaged? What brain circuitry is necessary for automatic speech? The basal ganglia, a system of subcortical brain structures—"subcortical" because they lie embedded beneath the cerebral cortex. They play a role in selecting appropriate motor actions by inhibiting ones you don't want to perform and are closely tied to emotion centers of the brain.[20] This patient lost the ability to blurt out emotionally charged idioms when brain damage compromised the functioning of his right-hemisphere basal ganglia.

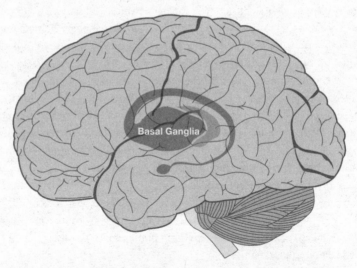

The basal ganglia, nestled underneath the cerebral cortex, are involved in selecting and inhibiting motor actions. Source: Modified from an image by John Henkel, of the Food and Drug Administration.

From a single case study, we wouldn't want to jump to the conclusion that the basal ganglia are always necessary for automatic speech. We also don't know what aspect of automatic speech they're responsible for or whether they're necessary for all types of automatic speech. But this one case suggests that automatic language—including spontaneous swearing—might be generated by brain circuits responsible for automatic processes other than language that are tightly linked to emotion centers. We'll explore each of these themes in the remainder of this chapter. But as an intermediate conclusion, we'll have to content ourselves with this: automatic swearing is localized differently from other types of language, and as a result it behaves differently when the system is stressed or damaged. Although specific brain areas like Wernicke's and Broca's are vital to using a lot of language, many others are also involved. Language is manifested heterogeneously in the brain.

$ % !

The evidence from automatic aphasia implicates the basal ganglia in automatic speech. On its own, this evidence would be suggestive but not conclusive. As it turns out though, the basal ganglia can come to have altered functioning in other ways. Tourette's syndrome is a hereditary neurological disorder affecting the basal ganglia,[21] which, in line with these structures' known functions, affects people's ability to control their own actions. Tourette's is characterized by the production of spontaneous and uncontrollable tics. These tics can take the form of movements of the body or face or undecipherable noises that sound like grunts or throat clearing. Or the tics can be actual words.

In some cases, the words that people with Tourette's uncontrollably utter are taboo. This symptom, possibly the best-known aspect of the syndrome, is called coprolalia (that's from the Greek *kopros* ["feces"] and *lalia* ["speech"]). Coprolalia is the feature of Tourette's focused on most by the media—see, for instance, the depiction of a person with Tourette's by Amy Poehler in the 1999 film *Deuce Bigalo: Male Gigolo*. The reasons for this fascination are perhaps obvious: losing control over taboo language is more shocking than losing control over clearing your throat. But despite this attention to the taboo side of the syndrome, only somewhere between 10 and 50 percent of people with Tourette's (depending on who's counting

and whom they count) actually display coprolalia.* Nevertheless, coprolalia has been documented in people with Tourette's speaking dozens of
languages, including English, Japanese, Czech, and even signed languages,
where signers uncontrollably produce obscene signs.† Some hearing people
with Tourette's also make obscene gestures uncontrollably (so-called
copropraxia).²²

For the people it affects, living with coprolalia can be challenging.
Imagine being unable to restrain yourself from uttering profanity in the
supermarket or the office or when picking your child up from kindergarten. Some people with coprolalia describe the compulsion as akin to
a sneeze—you feel a growing impulse that can only be alleviated by its release. Because it's hard to suppress coprolalia—just like sneezes—people
with this condition often take the path of avoidance, finding ways to stay
out of public as much as possible. They often also develop coping strategies for those occasions when other people are around. One of the most
effective is to replace words or parts of words when they burst out. Apparently, it's harder for many people with coprolalia to suppress or substitute
a whole word than to modify just the end of a word that's coming out any-
way. For example, if a person's impulse is to say *cock*, closing the mouth
intentionally toward the end of the word will produce *cop* instead. It might
seem strange to run around yelling *cop*, but it's hardly as socially stigmatized as the alternative.²³

If we set coprolalia and the swearing preserved in aphasia side by side,
we can see that they differ in revealing ways. For instance, nothing like the
substitution strategy has ever been described in aphasia, and aphasics don't
feel the welling up of an urge to curse experienced by most people with
coprolalia. But more interestingly, the specific taboo words selected are
mostly distinct. Aphasics with preserved automatic speech often produce
predominantly expletives expressing frustration or surprise, like *shit, fuck,*

---

* The difficulty in assessing the true incidence has several causes. Most critically, the diagnostic criteria for Tourette's syndrome have changed over the years, most notably in 2000, when
  the DSM-IV loosened them to include any person experiencing persistent vocal and motor
  tics (removing age-of-onset and frequency criteria, among others). This led to increased diag-
  nosis of Tourette's syndrome and a resulting decrease in coprolalia proportion estimates.
† I know what you're thinking. If Japanese doesn't have profanity, then how can a Japanese
  speaker have coprolalia? According to the three case studies I know of (summarized in Van
  Lancker and Cummings [1999]), Japanese coprolalia includes insults (like *baka* ["foolish"] or
  dobusu ["ugly"]) and childlike words for genitalia (like *chin-chin* ["penis"]). They're words
  that describe taboo concepts or that you wouldn't use in polite situations, even though the
  words themselves aren't profane.

or *goddamnit*. But the expletives present in coprolalia tend toward words for body parts and bodily effluvia, as well as racial, ethnic, and gender-based slurs. This is true across languages.[24]

Both types of profane words—those preserved in aphasia and those that burst out in coprolalia—express strong, transient emotional states. This has led some theorists to propose that swearing of both the aphasia and the coprolalia types is produced by different brain machinery than the rest of language.[25] As I mentioned earlier, it's possible that there's one pathway for producing a lot of language—the one that's been principally studied in humans and that in most people passes predominantly through the language centers of the left cerebral cortex and is used for the systematic, intentional composition of normal language. The second purported pathway is evolutionarily far older and shared with other mammals who themselves are bereft of anything like human language.

The limbic system, emotion-generating regions deep in the brain, dominate this proposed second circuit. The basal ganglia are directly adjacent to and tightly interconnected with brain structures that process emotions, like the anterior cingulate, the hippocampus, and the amygdala. These ancient brain structures appear to play a role in generating emotional states that create motor impulses, which the basal ganglia then have to regulate and selectively suppress. In the case of coprolalia, the compromised basal ganglia are unable to suppress verbal impulses along this pathway, which results in the characteristic expletives.

Work with other animals—particularly other primates—has revealed closely homologous circuits. For instance, when neurons in the limbic system of macaques or squirrel monkeys are stimulated, the animals spontaneously produce emotional vocalizations.[26] This implies that in the typical human brain, perhaps even mine and yours, everyday impulsive, automatic, emotional swearing may be driven by this very same circuitry—circuitry that is a mammalian or primate innovation rather than a uniquely human one. This circuit fills a vital evolutionary function for social beings, allowing an individual to transmit a signal identifying its internal emotional state readily and efficiently to conspecifics. If analogous circuitry is indeed responsible for reflexive human swearing, then it provides privileged access to emotion in the brain, laying bare a speaker's covert internal experiences unmediated by rational and deliberate planning.

But there's a caveat. This older, emotion-driven circuit doesn't behave the same way in humans as it does in other animals. As both Timothy Jay and Steven Pinker have pointed out, the vocalizations we produce when spontaneously swearing are conventionalized—they're the product of socially driven learning.[27] Swearwords are a different beast from shrieks or growls in that they have a specific learned form—you swear specifically in English or Chinese or ASL, whereas a monkey just shrieks in Monkey.

$ % !

The ramifications of these brain facts are manifold. First, it appears that the classic view of language as left-lateralized and cortical, subserved by a set of distinct and specialized brain circuits (e.g., Broca's and Wernicke's areas), tells only part of the story. We also use another pathway for language, built from machinery that's far older in evolutionary terms and shares little brain circuitry. Language is not monolithic in the brain.

And on a bit of reflection, this makes sense. We use language for multifarious purposes. We should hardly expect the brain systems that support the production of anodyne phrases articulating rational thoughts (like the present sentence) to have the capacity to do double duty, also connecting abrupt, hot, emotional states like frustration and rage to the spontaneous, primal utterances that express them.

And even this separation of the language capacity into two pathways surely understates the variety present in the brain. Although our evidence on the issue is poor at present, it wouldn't be shocking if we used distinct machinery for the various things we do with language. Do we use a separate pathway for conventionalized greetings (*Hi. How are you? Fine.*)? Do we use another for onomatopoeia—words like *cock-a-doodle-doo* that sound like what they mean? And is more detailed variation afoot even within the two ostensible pathways we've been discussing? Is our two-way distinction too coarse? Perhaps we recruit distinct pathways when we spontaneously express frustration versus anger, fear versus arousal. To date, we just don't know, in large part because the low-level neuroscience depends on having animal models to work with, and other animals share some but not all of our brain circuitry: Broca's and Wernicke's areas appear to be largely human specific, for instance. And other animals display some but not all of the communicative functions we deploy language for. In the future, better imaging techniques applied to the functioning human brain will surely reveal the extent of the diversity of neural instantiations of language behavior.

Finally, it's also worth reiterating here that it's impossible to equate different brain pathways with different words. We can produce the same words in different ways. When compromised brain function leads to language deficits, it's usually not specific words but specific ways of using them that are lost. Broca's aphasia usually impairs deliberate, intentional articulation of words and preserves some automatic speech. That is, even an aphasic priest who spontaneously and fluently produces "the most forceful oath of the tongue" when frustrated will be unable to intentionally articulate the very same word.[28]

5

The Day the Pope
Dropped the *C*-Bomb

By any account, Pope Francis has made interesting choices. He has foregone the traditional, opulent Papal Apartments, electing to reside in a small, modest bedroom in a Vatican guesthouse instead. He wears a silver ring instead of the traditional gold. And he has made a practice of washing feet each Easter—not the feet of priests as his predecessors did but those of patients at a home for the elderly and disabled, non-Catholics, and women. In aggregate, these many small acts of modesty have helped him build up a public image as the pope of the vulgar people.

Still, no one expected him to be quite this vulgar. On March 2, 2014, while delivering his weekly Vatican address, he slipped in a word that caught the world by surprise. He was speaking in Italian, and this is what he said: *in questo cazzo.* This translates literally as "in this dick," but since the offending word *cazzo* is used in Italian roughly as *fuck* or *fucking* are in English, in colloquial terms, he said something roughly equivalent to "in this fucking . . ." I'm no papal scholar, but I'm willing to go out on a limb and proffer that this is an uncommon turn of phrase for a pope, even one fresh off an appearance on the cover of *Rolling Stone.* The media ran with it—the story was featured on the *Huffington Post,*[1] NPR,[2] and the *Daily Mail,*[3] just to name a few.

This particular incident is so surprising and juicy in part because it runs afoul of how we expect the pope to express himself. Uttering a profane word like *cazzo* places him in an ideological double bind. If the curse word was accidental, then he's just as linguistically fallible as the next guy, which isn't necessarily the ideal public image for the professed terrestrial representative of God. Conversely, he might still be infallible, yet have intended to say *cazzo*. Again, likely not the image he means to project. All signs point to the former explanation—that this was a case of mistaken articulation. The clearest evidence is what he said next. Immediately after *in questo cazzo*, he corrected his phrasing to *in questo caso*, meaning "in this case," which seems more like something you'd hear from the Holy See. With a slip of the tongue, the pope revealed one more way that he's like the people. Or did he?

Everyone's tongue slips, including the tongues of people who are not the pope. Researchers who try to quantify this sort of thing report that people generate speech errors at an average rate of one or two errors every thousand words, or one error per ten minutes of speech.[4] But not all of these errors are equal. Some, especially errors that produce profanity, are particularly revealing. These are a big deal for the science of speech production—how people plan speech, select words to say, and articulate sounds. People produce profane and innocuous errors at different rates, which turns out to be one of the best ways to understand why we flub our words, how we're able to avoid errors, and how the brain manages all this. And as we'll see, in flubbing *caso*, the pope might have shown more of his hand than he intended.

$ % !

Producing language is one of the most complicated things you do, and yet you hardly ever notice you're doing it. When speaking fluently, you talk at a rate of about 120 to 180 words per minute.[5] That's a lot to churn out—two to three words per second. And despite the fact that words are separated by spaces in writing, you don't typically pause after each word (unless you're William Shatner).* This is in part because people in many speech communities feel pressure to keep talking once they've started. If, as a speaker, you don't manage to make sound come out of your mouth more or less continuously, someone else might believe that you have no more to say and jump at the opportunity to take the floor. Or they might become concerned that

---

* And if you, dear reader, actually are William Shatner, I take it all back! I'm a huge fan! I especially loved your Esperanto work in the cult-classic 1965 film *Incubus*! I have so many questions! Call me!

something is wrong and check to make sure you're not asphyxiating on your food or drifting to sleep. So you keep talking. And this taxes the system. In order to produce connected speech, you have to decide what the next word will be and start planning to say it before you're even done with the current word. You have to look ahead. This, in part, explains why speech is populated with *ums* and *likes* and other filler words that allow you to keep making sound, even if you don't know exactly what you want to say. And it also generates speech errors.

Among the different kinds of speech errors you make—swapping, dropping, and even adding sounds—certain errors most clearly stem from the preplanning of words that are a little farther down the assembly line. A simple but common type of error happens when a person anticipates a later sound and accidentally pronounces it too early. For instance, you might intend to say *shark pit* but instead accidentally mispronounce it as *park pit*. An error like this could only occur if the speaker were already planning later words (in this case, *pit*) while still articulating the current one (*shark*), because the *p* from the second word ends up at the beginning of the first word. It's also common to accidentally swap two sounds; you might know this as a spoonerism, but psycholinguists call it an "exchange error." A typical example would be intending to say *shark pit* but accidentally producing *park shit*. This again would arise because you were planning the next word while articulating the current one.

Because planning ahead causes exchange and anticipatory errors, we can actually harness them to reveal precisely how far ahead speakers plan. Some exchanges and anticipations are much more distant than just the adjacent word. We know this because linguist Victoria Fromkin of the University of California, Los Angeles, spent a large part of her career compiling a massive database of actual, observed speech errors. Among these errors were cases like *a Tanadian from Toronto* (instead of *Canadian*) and *Baris is the most beautiful city* (instead of *Paris*). These cases demonstrate how distant words can be and still exert influence on each other. *Canadian* and *Toronto* are two words and five syllables apart, while *Paris* and *beautiful* are four words and five syllables apart.[6] The distribution of speech errors suggests that when you make errors, you're already planning one to five words ahead of the word you're currently articulating.[7] And you're likely planning ahead even when you don't make errors.

So is it possible the pope's error was due to a mere planning inefficiency? Any single instance of a speech error could have many causes, and the only way we can tell that planning plays a role is in aggregate across many errors where the patterns reveal themselves statistically. But if we closely examine exactly what the pope said and what he was planning to say later, we can determine at the very least whether planning ahead is a plausible explanation for his *caso* to *cazzo* slip. So here's the full text of the sentence containing the error, as released by the Vatican in the official transcript of his prepared remarks. WARNING: The following may contain Italian.

Se ognuno di noi non accumula ricchezze soltanto per sé ma le mette al servizio degli altri, in questo caso la Provvidenza di Dio si rende visibile in questo gesto di solidarietà.[8]

In this critical sentence, the pope is making an appeal for charity, as you can tell from the English translation: "If each one of us does not amass riches only for oneself, but half for the service of others, in this case the providence of God will become visible through this gesture of solidarity."

Let's look at the words that followed *caso* for possible anticipation candidates. There's *la*, the feminine version of the definite article "the," and then *Provvidenza* ("providence"). Notice that *Provvidenza* has a *z* right before the last vowel, just as *caso* has an *s* right before its last vowel. Could Francis have been planning the long, complicated word *Provvidenza*, while still articulating *caso*, such that he replaced the *s* of *caso* with an anticipated *z*? It's possible. The two sounds are quite similar, and they're in the same place only two words apart, in words that are also of the same part of speech (which can increase the likelihood of errors).[9] But this close reading of the text doesn't lead to anything conclusive. At best, it tells us we should hold on to preplanning as a reasonable suspect.

There are, of course, other potential causes of the pope's error. The popular conception of speech errors holds that they're due more to meaning than mechanics. Sigmund Freud famously argued that when you make a speech error, it can reveal the inner workings of your unconscious mind.[10]

---

* But they're probably not the sounds you're thinking of. In Italian, the letter *s* in *caso* is pronounced how an English speaker would pronounce *z*, and the Italian *z* in *Provvidenza* is pronounced like an English *ts*. The double *zz* of *cazzo* is pronounced as an elongated *ts*. But this doesn't substantively change the argument I'm making; it just makes everything more complicated if you don't speak Italian, which makes it seem like the type of thing a considerate author would quarantine in a footnote.

inadvertently uttering *cazzo* ("dick"), could the pope have been talking about the virtues of charity but thinking about his vow of chastity? It's possible.

We've established that the pope might have erred due to the sequence of sounds he was planning to utter, and we've entertained the Freudian possibility that things underneath the vestments were on his mind. But he was also speaking under challenging conditions. This wasn't an oration he delivered alone in the shower. Public speaking can be distracting—feedback from the amplification system and things happening in the crowd draw your attention from the task at hand. And it can be stressful too. Both of these factors increase the rate of speech errors.[15] I have occasion to observe this frequently in daily life. I live in San Diego, and despite the reputation America's Finest City enjoys for outstanding news broadcasting,* the local announcers are a bit uneven. One particular afternoon host on the public radio station has a habit of stumbling over words, whether describing the upcoming segment on *All Things Considered* or announcing the names of companies providing local underwriting. And there have been some doozies, as I'm sure you can imagine, when the local businesses in question have names like *Chism Brothers Painting* and *Bastyr University*.† The pressures of public address must surely be as challenging for popes as for anyone.

What's more, the pope was speaking in a foreign language, which makes fluent speech harder. You make errors in a foreign language simply because you don't know the gender of a word or because you have a tenuous grasp of some detail of the language's grammar. (Do indirect object pronouns come before or after the verb? And for that matter, what's an indirect object pronoun again?) These lack-of-knowledge-based errors aside, your base rate of normal slips of the tongue also goes up—by 1400 percent.[16] So perhaps we shouldn't be surprised that delivering an address in Italian, instead of his native Argentinian Spanish, might ratchet up the frequency of a pope's errors.

---

* For a historical perspective, see the following documentary: Apatow, J., and McKay, A. (2004).

† Paradoxically, we hold people to a higher standard in exactly those conditions that are most likely to induce errors. Even knowing as I do that speech errors are an inevitable part of speech production, especially when a person is experiencing stress (as radio broadcasters probably do), I often find myself yelling at the radio, "Come on, step up your game! This is public radio! You don't think I donated just for the tote bag, do you?" And then I remember that I didn't donate this year, and I feel remorseful, but then I rationalize not donating with the thought that if the broadcasters didn't bungle their delivery all the time, maybe they'd deserve my money. The human mind is a silly place.

It seems reasonable that a fallible pope ought to be subject to the same pressures and linguistic traps as anyone else, and in this particular case they may have conspired to generate the linguistic C-bomb we witnessed. Indeed, with all these pressures at play, it's surprising that what comes out of your mouth—or the pope's—isn't just a stream of mistakes. Strangely, for the most part, it isn't. While we all make speech errors, the majority of words we produce really are precisely what we intend. And as we'll see, profanity provides the most revealing clues to how we accomplish this.

$ % !

Some psycholinguists have hypothesized that the only way you could possibly be as good at speaking as you are is by somehow monitoring your planned speech. You might, in this view, have an internal editor in your head that pays attention to the words you're planning to say, the order you're planning to say them in, and exactly how you're planning to pronounce them. It's like quality control at the end of the assembly line, right before the words get packed up and leave the factory. When your internal editor notices something about to go awry, it stops the conveyer belt and sends the offending word back for repair. Of course, some errors get through, so we know the editor can't be perfect, but the idea is that perhaps internal self-correction keeps your errors down to the acceptable level they're at.

How would we go about detecting such an internal editor? It's tricky because an editor would leave little trace. If there is indeed an editor, and if it's mostly successful, then there will be a few but not many errors to observe. Likewise, if there isn't an editor, there will be a few errors. The problem is figuring out how many errors a person would have made without an editor if we don't know whether there's an editor involved in the first place.

Let me flesh this out with our factory analogy. Suppose you want to know if a factory that cans diet soda has a quality control department. You can start by observing errors—every once in a while, someone finds a cockroach in a sealed can of diet soda. Let's say it's once out of every hundred million cans opened. Now, were thousands of other roaches in cans of soda caught by a vigilant quality control department? Or is the fabrication process itself so hygienic that an interloping roach finds itself trapped in a saccharine sarcophagus only one time in one hundred million? How would you tease apart these alternatives? Seems like a dead end (and not just for the cockroach).

But here's a possible way forward. To continue with the analogy, suppose you know that a cockroach finding its way into certain types of soda would be particularly devastating for the factory's reputation. For instance, suppose that the same factory packages the very same liquid not just as a brand-name soda but also, labeled differently, as an in-house, generic supermarket label. The only difference is the cans. And let's presume that the company has a greater incentive to ensure that the brand-name version is roach-free because it's a bigger source of revenue, and a single photo of just one roach in the brand-name soda will break the Internet and gut the company's bottom line. By contrast, the company might reason that people sort of half expect to find insects in in-house supermarket-label soda. Maybe that's even why they buy it. Here's the point. If there's no quality control department, then you'd expect to find cockroaches in the brand-name and generic sodas with about the same frequency; they're produced and canned in the same factory using the same process. But if you found far fewer canned roaches in the brand-name versus the generic soda, that would tell you that someone's making sure that when it matters more, mistakes don't make it out the door and onto the truck.

Several psycholinguists have used exactly the same logic with speech errors. To do this, you need to find certain speech errors that would have graver consequences than others. What's the linguistic equivalent of a cockroach in a brand-name can of soda? Well, the speech errors with the direst results are probably those that generate profanity. So the question becomes, when you put people in a position to say the wrong words, do they make the same number of errors, regardless of whether the error would produce profanity? If so, then there's no evidence of an internal editor. But if people make fewer profane mistakes than nonprofane mistakes, then that implies that people are internally suppressing the errors before they hit the tongue. They're self-monitoring language.

If you want to use this logic, you have to devise a way to induce speech errors in the lab. The first group to do this came up with a clever design.[17] You'll recall Michael Motley, the researcher with the provocatively clad research assistant in the Freudian slip study. He and his colleagues had people read carefully designed word pairs one at a time. Some of the pairs had the potential to become obscene if the participant made an exchange error. For example, *tool kits* seems totally innocuous, until you recognize how it would sound as a spoonerism. You can construct lots of potentially obscene lures like this: *bunt call, hit shed, duck fate, heap chore, fast luck,* and so on. The key question in Motley's studies wasn't whether you can get people to mistakenly mispronounce these. You can. Following the cockroach logic, the question was whether people would make fewer errors on profane-potential pairs like these than on pairs that do not threaten obscenity. So there was a second type of pair, like *tool kicks*. These words are identical in most every way to *tool kits*—they start with the same sounds, they're the same length when pronounced, they're the same parts of speech, and so on. The difference is that the errors you might make, producing things like *cool ticks*, aren't in the least taboo. Would there be more errors on these inoffensive pairs than on the offensive ones?

One methodological note, in case you're interested in trying this at home: Just reading pairs of words out loud doesn't yield many errors. So you need to boost the error signal. One thing you can do (and the researchers did) is stack word pairs one after another leading up to the critical one in order to set people up for failure. For instance, if you want people to make an exchange error on *tool kits*—swapping the initial *t* and *k*—you'd stack pairs in front of it as below. Try reading these aloud:

> *kind tiger*
> *calm time*
> *cold tea*
> *tool kits*

Setting people up with the swapped consonants before critical pairs like *tool kits* or *tin cable* makes participants much more likely to produce errors. This provides more opportunities for mistakes, which makes potential differences in the frequency of taboo and nontaboo errors easier to measure.

So, all other things being equal, do people make fewer errors on pairs like *tool kits*, where the result would be offensive, than on ones like *tin cable*, where it wouldn't be?

Two studies did this originally, in 1981 and 1982. The chart you see on the next page shows the number of errors people produced on average. There were many more neutral errors than taboo ones in both studies (though the difference was bigger in the second study). It follows that people in these studies were successfully avoiding errors specifically when the results would be obscene. Fewer roaches in the brand-name soda. People were self-monitoring.

But this is only the beginning of the evidence. If you're clever enough, as the researchers working on this are, you can come up with some other

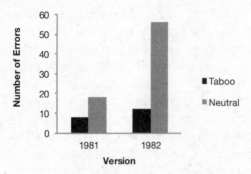

People make fewer speech errors when the result would be taboo.

things you'd expect to see if people were doing internal quality control. Here's one. Suppose you're on quality control at the factory and you find a can with an unwelcome passenger inside. Sending it back to be fixed or re-placed should add time to the process. So even when the ultimate product shows no sign of error, the time it takes to produce it could be a hallmark of monitoring. Mapping this over to speech, if the lower error rates with taboo word pairs are due to editing, then that should show up on how long it takes people to produce the pairs correctly—it should take people lon-ger to say pairs correctly when they've planned an error but subsequently taken the time to catch and correct it.

But people are constantly not making errors. We know that people make fewer errors producing taboo words, but that doesn't mean that every time they successfully avoid an error they're making a correction. Some proportion of the time, they're probably just getting it right from the out-set. We need a way to diagnose whether people are activating an internal plan to say something that they eventually don't produce. This is nearly impossible to do.

Except with taboo words. That's because even thinking about taboo words has special effects on the human body. When people say taboo words, their pores open up within seconds, and they sweat. And this is measurable.[18] Sweat conducts electricity, and the more sweat there is on your skin, the more conductive the surface of the skin will be. So, you can pass a very low level of electrical current across people's skin, say, on a fin-ger, and measure how conductive the skin is. When people start sweat-ing, the conductance increases. This is the basic logic behind what's known as the galvanic skin response (GSR). You might be familiar with this tool because it's one of the components of a traditional lie-detector test—skin conductance also changes as a function of anxiety, which may be driven by lying in some people.[19] Critically, although many words make people sweat, including emotional words like *murder* and *hate*, the most profuse and most reliable sweating comes from hearing taboo words.[20]

Suppose you were able to measure skin conductance as people performed the word-pair reading task. If merely planning a taboo word—even one that ultimately gets internally corrected and is therefore never pronounced—makes people sweat, then this should show up as an increase in skin conductance.

So here's the idea. You have people produce word pairs. Some, like *tool kits*, can induce profane errors, which, most of the time, people successfully avoid making. But looking just at the times when people didn't make an error, you can split these correct pairs into two groups. The first includes those where the participants' skin conductance spiked—suggesting that they had internally activated a taboo word, even though they didn't make an error. And the other group includes instances where there was no error and also no spike—suggesting that no profane word was ever even considered. And you look to see if the sweaty trials also take longer. Together, an increase in skin conductance and a longer time to produce the word pair would provide compelling evidence—albeit still circumstantial—that people were internally planning, but ultimately taking the time to correct, taboo words.

And that's what was found. The chart on the next page shows how long it took people to successfully avoid taboo errors when they had large and small GSRs.[21] As you can see, when they were sweating (the high-GSR group), they also waited significantly longer to start talking (the left bar is taller).

Here's another, lower-tech way to detect whether people are doing active editing before they speak. It has to do with the different types of errors people make. We've already talked about anticipatory errors (*hit shed* becomes *shit shed*) and exchange errors (*hit shed* becomes *shit head*). But people also make something known as a perseveratory error, which takes a sound pronounced earlier and repeats it later (*hit shed* becomes *hit head*). The taboo-error word pairs we've been talking about, like *hit shed* and *heap chore*, typically have the potential to produce an obscene word on either the first word or the second word but not both. As a result, for any given

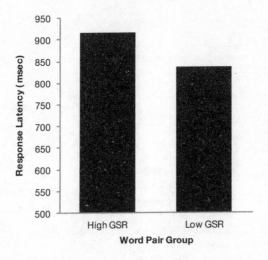

People who successfully avoid making taboo errors take longer to speak when they're sweating more (high GSR).

pair, an exchange error will always produce an obscene word (*hit shed* becomes *shit head*), and so will either an anticipatory error or a perseveratory error, but not both. In the case of *hit shed*, an anticipatory error produces the obscene *shit shed*, but a perseveratory error gives you *hit head*, which isn't so bad. With pairs in which the taboo word would be in second position, like *heap chore*, only the perseveratory error (*heap whore*) and not the anticipatory one (*cheap chore*) would be obscene. So if people are editing, then when the taboo words would be first, people should avoid the taboo word by making more perseveratory errors and fewer anticipatory ones. And when the taboo word would be second, this should reverse: people should make more anticipatory than perseveratory errors. That's what you see in the next chart.

What's more, you might notice that the numbers are far greater when the second word of the pair is taboo than when the first is—the light bar on the right towers over the others. Why would this be? Why would people make more anticipatory errors when the second word would be taboo? This too seems to implicate internal editing. If editing takes time, then presumably you'd be more likely to catch and stop an error when it's on the second word of a pair than when it's on the first. And if the errors you're most

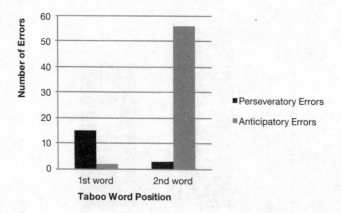

When people do make errors, they avoid generating a
taboo word.

vigilant about are taboo ones, then it follows that you'd be most likely to
catch and avoid those errors when they're planned for the second word.

This work, all conducted in the late 1970s and early 1980s, was influential at the time because taboo language provided a privileged way to detect
internal editing processes. More recently, researchers have been interested
in tracking down the brain basis for internal quality control of language.
And they've turned to the same foundational paradigm, with some major
technological additions.

The first new twist has people perform basically the same word-pair
reading task while tethered to an electroencephalogram (EEG) machine
that measures their brain waves.[22] In a nutshell, here's how EEG works.
Electrodes (lots of them—as many as 256 but more often 32 or 64) are applied harmlessly to the scalp. The electrodes measure fluctuations in the
electrical field, and the specific electrodes used in EEG experiments are
sensitive enough to measure microvolts—one-millionth the voltage of your
AA battery. A lot of things affect the electrical field measured by electrodes
placed on the scalp, including passing airplanes, elevators, and even the
muscles firing when a participant blinks. But it turns out that highly sensitive electrodes can detect something far more important to cognitive
scientists: the activity of neurons. When a neuron fires, at the chemical
level a bunch of ions flow into or out of it. And those ions carry electrical
charge (those pluses and minuses attached to Ca^{2+} or Cl^-). So when a nerve
cell fires, the flow of ions affects the electrical field around it. And when

thousands or millions of neurons oriented similarly and located close to each other fire at once, the electrical field change is strong enough for those sensitive scalp electrodes to measure.

Over many decades of research, neuroscientists have observed people performing hundreds of tasks while measuring their brain waves using EEG. And they've observed that certain types of behavior produce predictable changes in the measured electrical field. For instance, about four hundred milliseconds after you see a word, the electrical field centered over the top of your scalp deflects negatively. This is believed to index the process of interpreting the meaning of the word and integrating it into your ongoing understanding of the language you're reading or hearing.[23] Other components of the electrical signal relate to other specific behaviors and cognitive processes.

And so, when you wire people up to an EEG and have them read error-inducing word pairs, their brains produce different electrical signals, depending on what type of error they're avoiding.[24] A temptingly taboo pair like *bunt call* induces a stronger negative-going inflection over the center of the scalp about six hundred milliseconds after the prompt to speak, as compared with what happens when the same brain sees a neutral pair, like *bunt hall*. And this is true even when the person doesn't actually commit a verbal error. This tells us that the brain is doing different things when successfully avoiding a taboo error versus a neutral one. It doesn't reveal exactly what those different processes are—we only know for sure that neurons are firing differently—but it does tell us when that difference occurs. At six hundred milliseconds after the prompt to speak, people's brain activity diverges, depending on the type of error that would be produced. This suggests that the brain is doing something different when you're planning speech and not making taboo errors than when you're planning speech and not making mundane errors.

But we want to know not only whether something different is happening in people's brains and when but also what is happening. And because the human brain exhibits localization of function—as we saw in the last chapter, circuits located in different places execute different computations—knowing the location of the brain differences revealed by EEG could help us figure out what they mean. Unfortunately, it's notoriously challenging to extract locational information from EEG; changes to the electrical field measured at a particular electrode aren't necessarily due to the activity of neurons located directly below that electrode in the nearest piece of tissue. (The issue is complex, but it turns out that the direction the neurons are pointing matters, among other complicating factors.)*

But other techniques can tell us something about location. One is functional magnetic resonance imaging (fMRI), which measures fluctuations in the magnetic field from outside the body. When neurons fire, they use energy, and the more they fire, the more oxygenated blood flows to them, providing more energy (in the form of ATP, which you might remember from high school biology). Rushes of oxygenated blood can be measured by their magnetic signature, and so this can serve as a somewhat delayed and messy proxy for where neurons are firing in the brain. When you get more of this blood-flow signal to a particular region in the brain for one task than another, chances are the neurons in that region are doing more during the first task than during the second.

Applied to the same word-pair reading task we've been tracking, fMRI starts to fill in the picture of what the brain is up to while it's editing planned words. When you compare the brain's blood-flow signal during the taboo-eliciting pairs (*tool kits*) and neutral pairs (*tin cable*), you find that they're significantly different in one place, shown on the next page.[25]

The little blob in the lower part of the frontal lobe of the right hemisphere is in a region called the right inferior frontal gyrus, which is implicated in inhibitory control—your ability to stop or prevent yourself from doing something. Suppose you're waiting at a traffic light, for instance. It turns green, so you prepare to step on the gas, but then just as quickly, it turns red (perhaps because an emergency vehicle or train has overridden the usual light sequence). You need to quickly interrupt your plan to move forward. This appears to be the specialty of the right inferior frontal gyrus. It sends a hold-the-presses signal to stop action before it starts.[26]

Is it just me, or does that sound a lot like what an internal editor would be doing when confronting a taboo word that's about to come out of your mouth?

$ % !

---

* The real issue is that this is a type of "inverse problem." Here's the basic idea. Even if you know what the output of a complex system is, tracking down its causes turns out to be impossible. This is because although the output is determined by the system, the system is complex enough that many different system behaviors could produce the same output. For more, see Baillet, S. (2014).

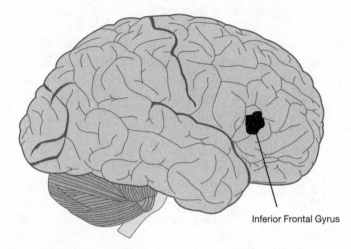

Inferior Frontal Gyrus

The right inferior frontal gyrus (a brain region involved in inhibitory control) experiences increased blood flow when people avoid taboo errors as compared with nontaboo errors.

The evidence from taboo speech errors and what happens when you avoid them implicates an internal process of self-monitoring. You are constantly censoring your words even before you articulate them in order to avoid slipups like the pope's. And it looks like, as far as the brain is concerned, suppressing an error—in particular an erroneous taboo word—is a lot like suppressing an action in response to an external stop signal.

But this is only the beginning. There are other ways to tell that people call in an all-systems-halt signal when they feel they're about to inadvertently say something obscene. We know this because psychologists have been trying to trick people into profane slipups in a variety of ways for decades.

One way is a well-known phenomenon called the Stroop effect. Basically, if you have people look at words and say what color font they're printed in, they do pretty well. Show them a word written in blue, and they can say it's blue. Unless, that is, the word printed in blue ink happens to be the word *red*. Then it gets a lot harder—people are slower and make more mistakes. That's the normal Stroop effect, and it's interesting in its own right because it reveals that you can't help but process the meanings of the words that you read, even when willing yourself to pay attention only to the color of the ink. You process meanings automatically and you're tempted to produce them. To avoid errors, you slow down.

Strangely, taboo words induce a Stroop effect as well: they interfere with people's ability to say what color a word is printed in. It's hard to demonstrate this on the printed page when you only have black ink (what century is it again?), but here's a quick-and-dirty approximation. We'll replace color with typography. Your job is to go through the list of words below, in order, and say whether each one is printed in italics, bold, or underline. Do it as quickly and accurately as possible. OK, go.

rib-eye
<u>stethoscope</u>
mountain
<u>pitchfork</u>
italics
donut
<u>library</u>
underline
fire-engine
barley
<u>philanthropy</u>
<u>bold</u>
stegosaurus
clandestine
cunt
fortuitous
bicycle
<u>fuck</u>
momentum

If all worked according to plan, you should have noticed that this task was harder for some words than others. The words that denote a particular font style, like the words *italics* and *bold*, should have been tough when they didn't match how they were printed. They should have taken longer, and you might even have made mistakes on them. That's the normal Stroop effect. But the taboo words should have taken longer as well: this is the taboo Stroop effect. You can see data from the first taboo Stroop experiment conducted on the next page, as described in a 1995 paper.[27] As you can see, the normal Stroop task causes a delay of about 150 milliseconds—compare

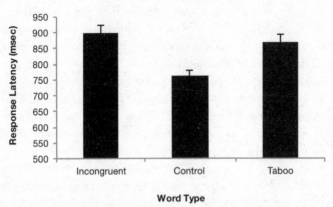

The normal Stroop effect causes people to name colors about 150 milliseconds slower, as shown by the difference between the control and incongruent conditions. The taboo Stroop is of similar magnitude, as demonstrated by the difference between the control and taboo conditions.

the middle bar with the leftmost "incongruent" bar. The taboo Stroop (the middle bar versus the rightmost bar) is nearly as large.

What causes the taboo Stroop? Certainly part of the story has to be the same as for the standard Stroop: it's hard to ignore the meaning of words that you focus on. Otherwise, you could selectively attend to the style of printing, and there would be no Stroop effect for us to talk about in the first place. Taboo words of course are particularly hard to ignore. But why do taboo words cause a delay in speaking, as compared with control words? Suppose, as we conjectured earlier, that when you perceive that you might mistakenly produce a taboo word, your internal monitor hits the brakes. This would lead to longer reaction times when the information that you're supposed to ignore (but might erroneously produce) is taboo. So the taboo Stroop effect could be explained once again by internal self-monitoring.

To be fair, there are other possible explanations. Perhaps the most convincing one sidesteps the issue of production and inhibition entirely. We know that seeing a taboo word evokes an emotional response. As I mentioned earlier, when you measure skin conductance as people simply listen to words, even when they don't need to speak at all, GSR is larger for taboo words than for neutral ones.[28] So an alternative explanation of the taboo Stroop effect is simply that this emotional response sucks up mental resources that you'd otherwise need to name the color of the word.[29] In effect, the emotional jolt you get from profanity overwhelms you to the point where other tasks you're trying to perform concurrently get put on the back burner and therefore take longer.

There's some corroborating evidence for this view. Experiencing strong emotions leads people to instantaneously encode a memory of what they're experiencing when that emotion hits, generating a so-called flashbulb memory, like the birth of a child, seeing the space shuttle explode, and so on. So if the taboo Stroop stems from a strong emotional response to seeing a taboo word in the middle of a psychology experiment, then your brain should encode an image of the event—in this case the word—that's stronger than memories typically encoded for less emotional experiences, like neutral words in the same experiment.

In fact, when you spring a pop quiz on people who have participated in one of these taboo Stroop experiments, they remember the taboo words they saw far better than they do neutral ones. Not only do they remember which words they saw, but they're also better able to remember what color the taboo words were printed in and even where they appeared on the screen.[30]

The upshot is that the taboo Stroop effect could provide further evidence that people are self-monitoring so they don't accidentally say the wrong word, but it's also consistent with this alternative, emotion-driven explanation.

There's another corroborating effect, similar to Stroop, which comes from a paradigm known as picture-word interference.[31] The basic idea is that you have people name pictures of familiar artifacts and organisms—hammers, tigers, and so on. This isn't hard. But the task gets slightly harder when you print words over the pictures. The words don't really interfere with your ability to perceive the object, but they can make it harder to name it, depending on how the word and picture are related. The details are tricky, but, generally, if you show people a word that sounds like the name of the picture (for instance, if you print "dock" over a picture of a dog), then people name the picture faster; however, if the word is related in meaning (like "cat" over a dog), people take longer.*

---

* A caveat in case you're a psycholinguist or plan to become one: the precise timing with which the picture and word appear on the screen affects the size of these effects.

In picture-word interference, a person names a picture as quickly as possible while attempting to ignore the written word. Taboo words interfere significantly with picture naming as compared with neutral words.

Just as there's a taboo version of the Stroop effect, there's a taboo version of this picture-word interference effect. If you print an unrelated, neutral word over a picture (above, left), people have no trouble naming the picture (it's a *dog*). But put a taboo word there (above, right), and people slow down, by about forty milliseconds on average.[32]

Again, an internal editor could be performing quality control here. Alternatively, this generalized slowing down could stem from the emotional reaction people have to the printed taboo word. Taboo words are special in several ways, which means that there are different ways that they could have the same effects on people trying to produce speech.

$ % !

Why did the pope blunder into profanity? For the same reasons the rest of us occasionally do. There's time pressure on a speaker to pronounce the present word while planning the ones to follow. Layer on top of that the stress of speaking in public, the challenge of negotiating a foreign language, and the Freudian attraction of thoughts you think you ought to suppress, and the pope's ability to say anything fluently is a small miracle (but not in the technical sense that would qualify him for canonization). Mistakes like the pope's reveal the pressures at work in every moment of language use. They underline what a remarkable feat we accomplish in navigating the gauntlet of potential gaffes at every turn of phrase.

And although we tend to notice speech errors when they generate taboo words—those errors really grind our ears—these are far less frequent than innocuous ones. We saw why. A person's internal monitor isn't as strongly compelled to censor unintended words when they're inoffensive. And so the particular error that the pope made (replacing *caso* with *cazzo*) is actually a bit surprising, not just because it reveals him to be fallible but because you'd think his monitor would kick in when it detected a potential profanity.

So why did the pope's self-monitoring fail in this particular case, spectacularly producing the Italian C-word? Possibly the pressures of public speaking in a foreign language overwhelmed his ability to self-monitor, and that profanity slipped by unchecked. But let me offer an alternate account. It's also possible that he slipped into a profane error because his self-monitoring system didn't know that he was about to make a profane blunder. In other words, the pope might just have revealed himself not to know the C-word in Italian. He may have produced a word whose meaning he didn't know and then corrected himself from the to-him-innocuous slipup. Ironically, by flubbing a profane word, he might have shown himself to be something less than the pope of the people that his public image suggests.

6

Fucking Grammar

In every language there's a logic to where words go. Nouns and verbs and prepositions snap into place to form phrases and sentences. This is grammar. I know that for some people, the mere mention of grammar triggers flashbacks to traumatic childhood moments, diagramming sentences on a chalkboard, covered in flop sweat, in front of a room of jeering classmates. But if you recognize yourself in this description, let me offer you some comfort. What a cognitive scientist like me means by grammar isn't the angst-inducing stuff of your childhood. We don't mean admonitions against double negatives or ending sentences with prepositions. These primary school lessons are called "prescriptive" rules of grammar. They're part of some authority figure's agenda about how a language ought to be used, rules laboriously hewn into young minds, where they are promptly forgotten.

That's not our game. A scientist's place is not to prescribe. The biologist's job, for instance, isn't to instruct the birds and the bees on optimal mating techniques. Nor is it the chemist's job to tutor the gasses on how to be noble. Scientists observe, document, describe, understand, and explain. Language science doesn't come down on one side or the other in debates about split infinitives or the Oxford comma (even though there's clearly a right answer).*

---

* The Oxford (or serial) comma is a comma inserted before *and* or *or* in lists of three or more items. For instance, it's the comma right before *and* in *I bought bread, milk, and carrots.* The main argument in favor of the Oxford comma is that it can help to resolve ambiguities. For instance, without the Oxford comma, the following sentence would be ambiguous: *This morning, the president met with lunatic fringe groups, the Republicans and the Democrats.* You could

Instead, it aims to describe and understand the language knowledge already teeming through the minds of people like you.

And a capacity for grammar is one of the most extraordinary things that evolution has imbued the human mind with. The rules of grammar that you know, implicitly, without any instruction and without ever reflecting on them, allow you to exercise the most powerful design feature of human language. You have the ability to string together new sequences of words to articulate any combination of thoughts you can come up with. And likewise, using your knowledge of grammar, you're able to understand any meaningful sentence a person might wish to assemble, no matter how unexpected.

This point, like many dealing with language, was perhaps best articulated by comedian George Carlin when he pointed out that there are certain sentences no one has ever said before and that therefore no one has ever heard, like "As soon as I put this hot poker in my ass, I'm going to chop my dick off." Or "Honey, let's sell the children, move to Zanzibar, and begin taking opium rectally."[1] The fact that Carlin could utter these specific sequences of words and that you could understand them is a testament to the combinatorial power of language, which the unique grammars of every language on earth provide.

All typically developing humans use grammar to combine old words in new ways. But other animals don't, at least not as powerfully and flexibly, and the ability to assemble and interpret previously unexperienced strings of words makes human language a qualitative leap beyond every other communication system in the natural world. So it's fair to say that grammar is kind of a big deal.

But what are the rules of grammar in your head like? That is, what is it that you know—and are able to deploy—in order to understand a sentence you've never seen or heard before, like *Honey, let's sell the children, move to Zanzibar, and begin taking opium rectally*? If you've never seen it before, the answer can't be that you memorized it. Instead, you must be able to see how the words fit together because they do so in systematic ways shared by other words in your language. Although you might never have used *the*

---

read this sentence as a list: the president met with three distinct groups. Or you could take the appositive interpretation, where *the Republicans and the Democrats* just provides more detail about the *lunatic fringe groups*. The Oxford comma makes it clear that you intend the list reading. The only argument against the Oxford comma is laziness. Really, dude, you can't type one more comma? It's right there by your middle finger. Yes, that finger.

children as the direct object of *sell*, you might have used *sell* with *the car* or *the house*. *Children, house,* and *car* behave similarly in English, and it's largely agreed that the rules of grammar in your head are general enough to cover a range of like words. Instead of knowing one rule for *car* and another for *children*, you probably know something about nouns in general and what you can do with them. Add to this your knowledge (again, implicit) that *children* and *car* and *house* are nouns, and you have the beginning of a story about how you can do new things with old words. You know, implicitly, general rules of grammar that you can apply to the tens of thousands of words you know to construct any one of millions—or, in principle, a potentially infinite number—of new sentences that no one has ever said or heard before.

But as soon as you add profanity to the mix, the rules start to change. Profanity, like the rest of language, follows the largely unstated and usually unnoticed but thoroughly essential rules of grammar floating around in your head. For example, a fluent English speaker might complain, "There's too much homework in this fucking class." I know this is grammatical—it's a sentence that English speakers produce and understand fluently—because I overheard and understood this very sentence when it was uttered by a real, live college student.* This sentence complies thoroughly with the general grammatical rules that American English speakers have in their heads.

But squeeze this sentence a little, and you'll find that its grammar is a little strange. And *fucking* causes all the trouble. Look at what happens when you substitute other words in place of *fucking*, adjectives like *stupid* or *inspiring*. On the surface, these seem like innocuous little changes that don't make much difference. But then again, swapping out a parachute for a tablecloth doesn't really make a noticeable difference until you jump out of a plane. So let's throw these sentences out of a plane—grammatically speaking.

We begin with *There's too much homework in this fucking class*, as compared with *There's too much homework in this stupid class*. To intensify exactly how stupid the class is, you can add *really* or *very* right before *stupid* to give you *There's too much homework in this very stupid class*. Admittedly, this sentence sounds a little clunky. But even if it won't win you a Pulitzer, it's still English. You can generally add adverbs like *very* ahead of

---

* Let's pretend that she wasn't talking about my class. And while we're engaging in self-delusion, why not also imagine that I've only heard this comment once.

adjectives like *stupid* without fear. Same with *inspiring*: *this very inspiring class*. But what happens when you try it with *fucking*? You get *There's too much homework in this very fucking class*. I don't know about you, but I just can't interpret this as English at all. It doesn't seem grammatical—it doesn't seem to me (or other native speakers I've asked) like a possible sentence in the language. In short, a sentence with profanity doesn't follow the same rules as those without.

Here's another stress test. Take a subtle variant of the same sentence. Suppose you put the word *fucking* not in the penultimate position but instead right after *too*, like this: *There's too fucking much homework in this class!* Most native speakers I ask agree that this is a possible grammatical sentence of English. Again, that doesn't mean that it's the type of sentence you'll see in a style guide or that your English teacher will make you recite. But remember, we're interested in what people actually say, not in what they're told to say (or not to say).

And notice what's special about this sentence. Other words, like *goddamn*, could replace *fucking*, as in *There's too goddamn much homework in this class!* Works fine. Same for *damn, bloody, darn,* and *friggin*. These all work. But all of a sudden, you can't replace *fucking* with known adjectives. It doesn't seem grammatical to say *There's too inspiring much homework in this class*. And it's not just adjectives: you can't replace *fucking* with any of the other types of words it commonly patterns with, like quantifiers (*some*) or intensifiers (*really*).[2] All of these give you clearly ungrammatical sentences, like *There is too some much homework in this class!* Apparently only profane words—or facsimiles of profane words, like *friggin*—can fit into this particular slot. This means that general rules of grammar can't account for this particular sentence pattern. Instead, you must know special rules of grammar that apply only to profanity and friends.

Importantly, you can't just toss *fucking* about willy-nilly. Hold out on the *fucking* in our sample sentence until the very end, and you get *There is too much homework in this class fucking*, which sounds pretty ungrammatical. So the rules you have internalized about *fucking* aren't just more lax—you know precisely where *fucking* can go, what it can go with, and, importantly, where it can't go. And what's most fascinating about these special rules is that you know them implicitly. Before reading this chapter, you couldn't have learned them through explicit instruction—I'd bet my shirt that no one ever sat you down to explain that *fucking* can go not only before a noun but also after a quantifier. You learned these rules through observation, induction, imitation, and trial and error. You know a lot about the grammar of *fucking* without ever having brought it to conscious awareness.

Here's another example to tease your intuition. When you stick *not* into a sentence, you negate some part of its meaning. Compare *Let's sell the children* with *Let's not sell the children*. Important distinction. But this general rule meets its match when confronted with certain profane sentences. For example, compare *You know jack-shit* and *You don't know jack-shit*. Is there any difference at all? Most people agree that both versions mean that you know nothing. So how can it be that *not* has no effect? Could we once more be in the presence of a special rule for profanity?

For a science of language, special rules like these are both an abomination and an invitation. They call into question the very property that makes human language so expressive and powerful: the ability to flexibly mix and match words with general rules. And yet, as our job is to describe and understand, we're forced to confront them. What is the nature of these rules? What do you know about the grammar of swearing in your language? And why do we have these rules in the first place? These questions have driven language scientists to sully themselves a bit by digging into the nitty-gritty and exceptional details of dirty grammar. And so I invite you to reflect for perhaps the first but hopefully not the last time on some of the things you know—but don't know you know—about the profane grammar of your language.

$ % !

Profane words in English occupy nearly every grammatical category. There are, of course, the familiar verbs (*to fuck*, *to shit*), nouns (*a fuck*, *a shit*), and so on. But one of the most grammatically bizarre is in evidence in sentences like the one I just mentioned: *You don't know jack-shit*. Here, *jack-shit* is acting as something called a "minimizer." You've probably never heard of such a thing, but still you use them all the time. If you had to come up with a nonprofane replacement, it would probably be the more frequent *anything*—that would give you *You don't know anything*. In English, we have a variety of minimizers like *anything* that go along with negation and serve to emphasize how complete that negation is. These are words (or expressions) like *at all*, *one bit*, or *a drop*, all of which can be appended to negated sentences like *You don't know* or *He doesn't drink*. Some of these minimizing words are general, like *at all*, which you can stick into pretty much any negation: *I do not skydive at all* or *It doesn't hurt at all*. Others are quite specific to the action, like *a morsel* in *He didn't eat a morsel* or *a drop*, which appears only to be not spilled or not drunk.

Minimizers like these follow a general rule: they have to be used in a sentence that describes something nonfactual—a sentence stating explicitly that something isn't the case. Negation with *not* is of course the best way to do this. So *He doesn't skydive* provides a nice nonfactual home for *at all*. But a question is also somewhat nonfactual, because it questions whether something might be true. So you can stick *at all* and other minimizers into questions, like *Do you skydive at all?* Expressions of doubt work the same way because they also allow a shadow of possible nonfactuality to creep in: *I doubt you skydive at all*. And so on. But as soon as your sentence becomes factual—as soon as it asserts something as true—then you can't use *at all* or its ilk. It wouldn't make sense to say *I skydive at all* or *He really wants to skydive at all*. That's a general grammatical rule about words like these. They're only viable when the factuality of the statement is put linguistically in doubt.

But *jack-shit* and its profane peers flout the rule. You can say *You don't know jack-shit*, using it in a negative context, but you can also just as easily say *You know jack-shit*. Same with *dick*—*I don't draw dick unless the price is right* is fine, as is *I draw dick unless the price is right*. And other terms like *crap*, *shit*, *fuck-all*, and the like all behave the same way. For instance, *You get fuck-all until you say "please"* works just as well as the negated version.* This special set of profane words and their strange behavior in this context have gained such notoriety among syntacticians (people who study

---

* As I said, this is exceptional behavior. You can say *I don't know anyone* but not *I know anyone*. But there are a few other words that seem to behave something like the profane ones, *dick* and others, in that they can be used in both positive and negative contexts. And one of these, occasionally, is *anymore*. Most Americans are perfectly willing to say *Honey badger doesn't give a shit anymore*. But what about *Honey badger gives a shit anymore*? Fascinatingly, there are actually regional differences on this. If you think that it sounds perfectly fine to say *Honey badger gives a shit anymore*, then you (or your parents) are probably from the Midwest, particularly Ohio and Pennsylvania. You have so-called positive polarity *anymore*—you can use it, just like *dick*, in the presence or the absence of negation. The rest of the country, and possibly the world, will strenuously object to this use, largely because they categorize *anymore* as requiring a negative context. The difference between positive polarity *anymore* and vulgar minimizers like *dick* or *jack-shit* is that sentences using *anymore* have different meanings when negated and not—*I go there anymore* and *I don't go there anymore* mean different things, whereas *You know jack-shit* and *You don't know jack-shit* mean about the same thing.

grammar for a living—yes, this exists) that they've been given a name. Two names, actually. They're sometimes called "vulgar minimizers," which is an apt description because they're both vulgar and they minimize what precedes them.[3] But "vulgar minimizers" isn't as evocative as their other name, "squatitives," in honor of the special place of the word *squat* among them.[4]

More than their relative flexibility, the really remarkable thing about squatitives like *jack-shit* is, of course, that the negative and the positive versions of the sentences seem to mean roughly the same thing. As I mentioned earlier, this really is quite strange because putting a *not* in a sentence usually reverses some component of the meaning. *Let's sell the children* should mean roughly the opposite of *Let's not sell the children*. So when you use profane words as minimizers, the affirmative and negated versions of the sentence are similar in meaning. *You don't know dick* is roughly synonymous with *You know dick*. *He doesn't know jack-shit* means the same thing as *He knows jack-shit*. It's almost like we're looking at the sentence version of the English *flammable-inflammable* mess.*

To a first approximation, these profane words seem to be subject to special rules that simply don't apply to the rest of the language. They're outliers, but not random ones. They form little coalitions that pattern alike among themselves but flout the rules that apply to nonprofane words.

$ % !

In Emma Lazarus's poem "The New Colossus," which is inscribed on the Statue of Liberty, you'll find the famous verse "Give me your tired, your poor, your huddled masses yearning to breathe free." While this particular sentence has had immeasurable social impact, linguistically speaking it's grammatically unremarkable. The sentence is all constructed around the verb *give*, and in this sentence, it's doing what it normally does. The sentence explicitly identifies both the things that are to be given (*your tired, your poor*, and so on) and also the recipient, *me*. Who *me* is, I suppose, is

---

* To be clear, I'm not talking about irony here. It's always possible for a speaker (or writer) to say anything while really meaning the reverse. For example, you can say *Mary doesn't know jack-shit* ironically to mean not what it literally means—that she doesn't know anything—but instead to mean the reverse. For instance, *Mary has spent twenty years as a veterinarian caring for orphaned kittens and puppies, so obviously she doesn't know jack-shit about animals.* *Jack-shit* and other squatitives make a sentence and its negated opposite unironically mean the same thing.

subject to interpretation—it could be Lady Liberty or more likely the nation she represents—but it's important for *give* to have a recipient. It's kind of the sine qua non of giving. You can't give something without giving it *to* someone. Consequently, even when not explicitly stated in a sentence, the recipient of *give* is almost always implied and inferable from context. For example, the statement *I don't give handouts* implies that there's someone you don't give handouts *to*. Of course, you can put this person in the sentence: *I don't give handouts to bums like you, Mr. Lebowski.* But even when such a statement does not expressly identify the recipient, it goes without saying that someone is or isn't getting something.

But profanity again is the exception. When you give profanely, and here I'm thinking specifically of *giving a fuck*, the rule about *give* having a recipient doesn't appear to apply. You can *give* (or choose not to give) *a fuck*—without any potential recipient in mind. And the same goes for *a shit*, and *a damn*, and so on. For example, the famously resilient honey badger can reportedly be stung by a thousand bees, with what consequence? *He doesn't give a shit.* At the end of *Gone with the Wind*, Scarlett O'Hara asks Rhett Butler what she should do when he leaves. His answer: *Frankly, my dear, I don't give a damn.* This kind of giving or nongiving—of *shits*, *damns*, and *fucks*—is grammatically special. You don't mention the recipient: you don't specify whom you don't give a damn to. But it's even more bizarre than this. Not only is there no explicitly mentioned recipient; there's not even an implied one. We can tell for sure that there's no implied recipient because you couldn't even force a recipient into such sentences if you had to. It doesn't make sense to say *I don't give you a fuck* or *I don't give any fucks to you* to mean *I don't care.*

So why can't you *give a fuck* to anyone? One reasonable explanation could be that this is merely a consequence of *I don't give a fuck* being a fixed expression. Maybe it's a set of words that implies a recipient but into which you can't force one because those five words have to be said in exactly that order, as though *I don't give a fuck* were one single word spelled with internal spaces. The problem with this argument is that the words in *give a fuck* are in fact quite flexible. You can make it passive: *No fucks were given.* Or you can be crystal clear: *Not one single fuck was given.* And you can modify the *fuck* that you're not giving: it can be a *flying fuck*, the *slightest fuck*, or a *single fuck*. No, it's not that *give a fuck* is too rigid to admit a recipient. It's that there's something about recipients that *give a fuck* doesn't like.

You're going to start detecting a trend here, because it seems, again, that there's a special grammatical rule at play for *give a fuck*, one that also applies to *give a shit*, *give a damn*, and so on, but doesn't pertain outside of the realm of profanity. It's not that *give a fuck* is more lax, as in the case of *fuck-all* and other squatitives. No, in this case, the grammar is actually more rigid for *give a fuck* than for *giving* anything else. The general characterization of *give* and how it works (it has an explicit or implicit recipient) doesn't apply equally to all its uses. In order to use these profane expressions grammatically, you must know very specific things about how to make a sentence—grammatical patterns specific to particular uses of selected words.

$ % !

So we've now seen that in some cases profane grammar is more flexible, and in other cases it's less flexible than the grammar of the language as a whole. But on the whole, the differences we've seen have been relatively superficial—subtle changes in specific ways that words can or cannot be used. How deep does the special behavior of profanity go? Are there ways in which profanity seems to follow its own, qualitatively different system of rules entirely? Maybe.

It's generally agreed in grammar circles that every sentence has to have a subject. In English, you usually express the subject overtly. For example, look at the sentences in this paragraph. The first sentence has the subject *it*, which the verb *is* (contracted to *'s*) agrees with. The next sentence has the subject *you*, which *express* agrees with. Now, sometimes a sentence has no overt subject. Imperatives are an example of this. In *Look at the sentences in this paragraph*, there's no subject. But still we all know who's doing the looking: the person to whom the imperative is directed. You, dear reader, are the subject of *Look at the sentences in this paragraph*. Imperative sentences like this still have a subject; it's just implicit.

The idea that subjects can be implicit is a neat notion because it allows us to preserve the generalization about English sentences that they all have subjects. Some are overt; others are implicit. That's believed to be a general rule of English. A very general rule. Science likes general rules because generalizations enable concise descriptions and explanations of diverse observations. Gravity explains both orbiting planets and plummeting skydivers, and that's a good thing.

So suppose all sentences have subjects. Great. In that case, what's the subject of *Fuck you*?[5]* It's not obvious. You might be tempted to think that the *you* in the sentence is the subject. And certainly in the case of the similar sentence *You fuck* (a declaration of what *you* do), the subject—the one performing the action—is obviously *you*. But in *Fuck you*, the *you* can't be the subject because *you* isn't performing an action.

In that case, you might reasonably conjecture, *Fuck you* is probably an imperative. And if it's an imperative, it has an implicit subject, just like *Look at the sentences*. But that can't be true either. And we can tell for a very subtle grammatical reason that I'll now attempt to explain.

Syntacticians pay close attention to how words can or can't combine in order to figure out what's really going on under the surface—in this case, whether something is a subject or not. Here's a clever test we can use. When the subject and the object of a verb refer to the same person or thing, something special happens. The object adds *-self* to the end. For example, if I want to describe an act in which you cleaned yourself, I couldn't say *You cleaned you*; I'd have to say *You cleaned yourself*, just like *I cleaned myself*, *He cleaned himself*, and so on. So we know that whenever we see these reflexive *-self* pronouns, the subject and the object are the same person. This is a kind of grammatical test you can apply to sentences.

The powerful thing about this test is that it also detects the implicit subjects of imperatives. If I wanted to tell you to clean yourself, then I would say *Clean yourself*, not *Clean you*. Because *yourself* is required, we know that the implicit subject of an imperative must be *you*. Neat. Further evidence that imperatives have *you* as an implicit subject. But notice what that means about *Fuck you*. If it were *Fuck yourself*, then we'd know that this was an imperative with *you* as the implicit subject—just like *Clean yourself*. And indeed, it's possible to say *Fuck yourself*, but this means something different from *Fuck you*. *Fuck yourself* is an actual imperative—it's a command for the subject, *you*, to perform an action, *fuck*, on an object,

---

* This is the topic of a classic piece of scholarship by James McCawley, a former University of Chicago linguist whose PhD from MIT was supervised by Noam Chomsky. By all accounts, McCawley was a polymath (for instance, he had several degrees in math), a prodigy (who started as a student at the University of Chicago at sixteen), and an inveterate prankster. Under the pseudonym of Quang Phuc Dong, ostensibly of the South Hanoi Institute of Technology (or SHIT), he wrote several seminal papers in what he called "scatolinguistics." The first, "English Sentences Without Overt Grammatical Subject," deals with the grammar of *Fuck you*. McCawley died in 1999 and with him a lot of the fun of linguistics.

which is also *you*. But with *Fuck you*, the subject can't be *you* because the object is *you*, not *yourself*. The subject has to be something or someone else.

So maybe *Fuck you* is just a special kind of imperative with a subject that's not *you*. We can put *Fuck you* to a number of other grammar tests to diagnose whether it's an imperative. And they all come back negative. For example, you can negate imperatives—for instance, *Don't read this sentence!* But you can't say *Don't fuck you!* You can add *please* or *do* to the front of imperatives: *Please read this sentence. Do look at this sentence.* But there's no way to interpret *Please fuck you* or *Do fuck you*. By all measures, *Fuck you* is not an imperative, and if it's not an imperative, it doesn't have an implicit subject, and because it also doesn't have an overt subject, that means it has no subject at all.

It's not just *Fuck you* that's missing a subject. Other vulgar maledictions are in the same boat. *Damn you* works the same way. Notice the same difference between *Damn you* and *Damn yourself* that we saw before. *Damn you* isn't telling you to perform an act of damning on yourself, but *Damn yourself* is. And again you can't negate it to make *Don't damn you*. Same with *Screw you*. It appears that *Fuck you*, *Damn you*, and *Screw you* aren't imperatives. And as a result, none of them have a subject, not even an implicit one.

Right now, you might be thinking about God. As a subject, I mean. Couldn't *Damn you* really be a shortened version of *God damn you* or *May God damn you*? And likewise for *Fuck you*, couldn't it really be *May God fuck you*? It's possible—at least for *Damn you*—that this is the historical source of the expression, as evidenced by the presence of *God* in *goddamnit*. But looking just at the grammar of the language as it's used today, there's no *God* left in *Damn you* or *Fuck you*, and we can tell by using the same reflexive pronoun test that showed us that *you* isn't their subject. Suppose you want to denigrate not the person you're talking to but some third party. You'd say *Damn him* or *Fuck her*. Well, it turns out that if the person you want to denigrate isn't a person but a deity, then you can perfectly grammatically (albeit blasphemously) utter *Fuck God* or *Damn God*. And here's the rub. If *God* is the subject of these sentences, then we shouldn't be able to say *Fuck God*. It would have to be *Fuck himself*; God is the subject, so the direct object should agree with it. But you can't say *Fuck himself* to mean *Fuck God*. And that implies that *God* is not the implicit subject of either *Fuck God* or *Fuck you*. They don't appear to have any subject at all.

This is a big problem. These profane maledictions are breaking arguably the most important rule of grammar. All sentences are supposed to have subjects, whether overt or implicit. That was the laudable generalization we started with. It's as if we've found one type of matter that the rules of gravity don't apply to. And so one of two conclusions follow.

One: *Fuck you* doesn't have a subject. But it's grammatical. And if all grammatical sequences of words are sentences, then we have to conclude that some sentences, like *Fuck you*, can live without subjects. That's going to be a hard pill to swallow. There's an exception to gravity.

If you don't like that, you do have another option. Conclusion two: all sentences still have subjects. But *Fuck you* and other maledictions are something other than sentences. The reasoning behind this would be the following syllogism: Sentences have to have subjects. *Fuck you* doesn't have a subject. Therefore, *Fuck you* is not a sentence. By this logic, sentences make up only one of several types of things you know how to say in English. There are also other things, like expletives. Perhaps certain expletives follow their own, distinct rules of grammar. Sentences have subjects. Expletives need not. They're a whole separate class of things people know how to say. This would be as big a deal for linguists as finding a type of matter that's immune to gravity would be for physicists or discovering a new phylogenetic kingdom would be for biologists.

And it's not just *Fuck you*. When you start to dig, you find that other profanity places you astride the horns of this same dilemma. Consider, for example, an utterance like *White wedding, my ass!* Is this a sentence? To begin with, it's not clear what the subject is here. It might be *white wedding*, or it might be *my ass*. Or neither. But that's not the real problem. Something else is missing. If you look closely, you'll see that there's no verb. And it's not like there's an implied verb. What could the verb possibly be? You couldn't say *White wedding is my ass!* Sentences need not just subjects but also verbs. If this is a sentence, it's profoundly degenerate.

You can see the problem even in one-word utterances, like the isolated word *Fuck!* There's one way to use this word that can, in fact, form a real sentence: an imperative one in which *you* are the implicit subject of a commanded action. For instance, it might be a command a breeder gives to her goldendoodles when they're in heat. *Fuck!* But the more common way to use the same single word *Fuck!* does not form a normal sentence. When used as an expression of frustration, anger, or excitement, it has no subject. No one is being instructed to do anything to anyone else. The same ambiguity between sentence and expletive is present in *Shit!* or *Crap!* or any expletive that also happens to be a possible verb. Expletives appear to have their own rules of grammar.

And although these utterances might not be sentences that we could construct using the general rules of grammar we've reviewed so far, they are still subject to very precise grammatical constraints. For instance, consider the nuances surrounding *White wedding, my ass!* For one thing, you don't have much leeway with whose *ass* it is—you couldn't get away with saying *White wedding, his ass!* or *White wedding, our asses!* And it seems like it has to be the word *ass* or a near synonym in that last position. So you could say *White wedding, my tuchus!* or *White wedding, my butt!* But it would be harder (though possibly still acceptable) to use other parts of the body: *White wedding, my hymen!*

The upshot is this: Certain types of profanity, from *Fuck you* onward, belong to their own class, or classes, of utterance. They're not sentences by any normal definition; nor are they abbreviations of full sentences that omit little bits. They're their own class of thing that you can utter. There's a chasm between the grammar of profanity and that of the language as a whole.

$ % !

And yet, despite this profound specialness, the profane utterances we've been looking at—even though they aren't normal sentences—still follow some general grammatical rules. For instance, in *White wedding, my ass*–type sentences, even though the pronoun pretty much has to be *my* and the noun has to be a posterior-related body part, there's nevertheless some flexibility. I believe it's still grammatical to say *White wedding, my fucking ass!* or *White wedding, my big fat Greek ass!* That is, you can use the very same normal rules for putting things together into sentences, in these cases modifying nouns with adverbs, adjectives, and the like, that apply in the language in general. So these utterances live in a nebulous space. On the one hand, they're a totally different type of thing—not like any sentence we know of. On the other, they can hook into the language's general rules of grammar in limited ways. Profanity has its own grammar, but it is built on top of the general principles that govern the language as a whole.

Here's another case of a specific grammatical pattern that still follows other general rules. The verb *tear* is usually transitive, meaning that it has a direct object. For instance, you might say *I tore my hamstring*. Here, *I* is the subject, the "tearer," and *my hamstring* is the object, the thing affected by the tearing, or the "torn," as it were. But sometimes, rarely, *tear* can have more than one object. It can be "ditransitive." An example of ditransitive verb use is *Mary tore me a new asshole*. There are two grammatical objects, *me* and *a new asshole*.

Just like with *give a fuck*, there's a little slack in this pattern. And this is where the rest of what you know about the grammar of your language comes in. The verb doesn't have to be *tear*. You can also *rip, ream, pound*, or possibly even *fuck someone a new asshole*. And there's a little leeway with the *new asshole* as well. It can be *a new one, another asshole*, or really anything that describes a new orifice. If I'm not mistaken, then, it would fit the pattern to say that you're going to *shag someone a supplementary shit shoot* or *hammer him home a hasty Hershey highway*. Like *give a fuck*, this grammatical pattern imposes constraints on what can occur in it, within limits. But otherwise it behaves as you would expect, given the rest of the language.

So why do these profane grammatical patterns flout certain grammatical conventions of English while obeying others? In some cases, it's hard to know. But perhaps not in all. There might be a hint of reason in this last pattern we looked at: *tear someone a new asshole*. Unlike general rules of grammar (the rule that sentences have subjects, for example), the patterns we've been looking at encode a very particular meaning or are tailored to a specific function. The *tear him a new one* pattern conveys the particular meaning that the person in the subject beat up the first object physically or verbally. And this meaning might explain how it patterns. Not just any verb can occur here—only verbs that can plausibly describe an act of orifice-creation qualify. And the verb takes not just any direct object. The first one has to be someone or something that can be beat up, and the second has to describe a new orifice. Could meaning or function impose constraints on the grammatical behavior of profane language?

This might be clearer in a different pattern of profane grammar. Consider where you can stick *the fuck*, specifically, when used in questions. *The fuck* can of course be inserted directly after *what* to make *what the fuck*. Certain other "*wh*-question words," like *who* and *why*, work the same way: *Who the fuck do you think you are? Why the fuck would I tell you?**

---

* Much of this discussion is inspired by Fillmore, C. J. (1985). We miss you, Chuck.

But the king of these *wh*-question words really is *what*—the now pervasive acronym *WTF* usually refers to *what the fuck* rather than *why* or *who the fuck*. (Notice that the list of relevant *wh*-question words includes *how*, even though it doesn't start with a *wh*-, as in *How the fuck should I know?* But it might not include *which*—many, but not all, English speakers find it ungrammatical to ask *Which the fuck should I choose?*) When used in this way, *the fuck* is largely interchangeable with *the hell, the shit, the devil, the deuce*, and a few others, with corresponding changes in intensity.

This is yet another case where profanity is behaving differently from the rest of the language. You can insert *the fuck* and friends into *wh*-questions, but only *wh*-questions of a certain type. The rule appears to be that the *wh*-word has to be the very first word of the clause. So it's grammatical to say *What the fuck did you open that jar with?* but not *With what the fuck did you open that jar?* You also can't say *You opened the jar with what the fuck?*

These inserted *fucks* are also unique in the language because, as with the other profane patterns we saw before, there are no other words you can drop into just these places with these precise restrictions. For instance, it's possible to insert *did you say* into a question directly after the *wh*-word to ask for clarification, as in *What did you say you opened that jar with?* But in this case, it's also totally acceptable to put the *with* at the front, as in *With what did you say you opened that jar?* or even to embed it: *You opened the jar with what did you say?* In other words, the rules for profanity are similar to but different from those for the rest of the language. This should seem quite familiar.

But this pattern is also revealing for the question at hand: Is the grammatical behavior of this pattern constrained by the meaning or function of these words?

Consider the following facts: You can embed a *wh*-clause in a larger sentence. A clause is just a sentence-like thing inside another sentence. For instance, a sentence like *I can't imagine what he cooked* contains the clause *what he cooked*. Now, in this case, *the fuck* can be inserted just fine, right after the *what*, because *what* is still the beginning of the embedded clause. This gives you the fully grammatical *I can't imagine what the fuck he cooked*. So far so good. But you can't always stick *the fuck* there. Sentences that seem superficially quite similar do not seem grammatical to most people. For example, what do you think about *I can't disclose what the fuck he cooked*. If you agree that this sentence seems strange, or at least stranger than *I can't imagine what the fuck he cooked*, then this must be due to a difference between *imagine* and *disclose*. Why is *imagine* more welcoming to *the fuck* than *disclose* is?

They mean different things. In the case of *imagine what he cooked*, which can have *the fuck* in it, the thing that he cooked is unknown, wondered about. In the case of *disclose what he cooked*, the thing that he cooked is some specific thing that the speaker knows but doesn't want to reveal. So could it be that sentences that you can embed *the fuck* in express uncertainty about the event? Let's see. You can say *I have no idea what the fuck he cooked*. And sure enough, it expresses uncertainty and allows *the fuck*. But when the sentence expresses certainty, then all of a sudden *the fuck* seems out of place: *I can't eat what the fuck he cooked* sounds strange to many people, as does *This is what the fuck he cooked*. In other words, the grammar of *the fuck* is in part constrained by what the sentence means. You insert *the fuck* after *what* to express incredulity at something unknown. And as a result, sentences that don't express uncertainty don't allow you to insert *the fuck*. This shows that grammar cares about meaning. What you know about how to put words together is sensitive to the meaningful work you're trying to do with grammar.*

---

* One final thing that's interesting about this case is that the profane word (*fuck* or *hell*) looks like a noun—it follows *the*, as nouns are wont to do—but it doesn't behave like just any noun. That is, if we care about what people know about the grammar of their language, part of what we want to know is what categories they're using in their grammatical rules. For instance, English appears to have a rule in which *the* can precede nouns in general, from *aardvark* to *zythology*. (*Zythology* is of course the study of beer making.) But the rule of grammar that allows us to create and understand *what the fuck* is far less general. You can't insert just any noun whatsoever after a *wh*-word: *What the aardvark is on your plate?* seems out of place, as does *How the zythology am I supposed to drink this?* So how can we describe what people know about *the fuck* or *the hell*? What rule of grammar allows these words to be inserted as we've seen they can be? It certainly can't be a rule of grammar stating that nouns in general can be inserted after a *wh*-question and *the*. It must be more specific. It must say that a short list of specific nouns are available for this particular rule.

So you might again be tempted to think that *the hell* or *the fuck* is just a fixed expression or "idiom," or maybe that *What the hell* is the idiom. If so, then you just memorize the whole thing and forget about applying rules. *What the hell* is basically just a big word with some spaces in it. But the problem is that *hell* and *fuck* in this expression are variable—they act a lot like any regular, lively noun. For instance, they're available for certain normal grammatical operations you perform on nouns in general. You can modify nouns by putting adjectives in front of them. That works here too: *What the bloody hell?* You can add an adverb then an adjective, as in *What the everlasting bloody hell?* The point is that *hell* seems to be acting a lot like a noun here, just with very limited flexibility. It has to be preceded by *the* and not, for example, *a. What a hell!* doesn't cut it. And it has to be singular, as there's no sense to be made from *What the hells!*

$ % !

There's another reason why some profane patterns follow or refuse to follow just the rules they do. And that has to do with their history. To see this, let's look at two other uses for *the fuck*.[6] Although they look quite similar, each has its own idiosyncratic meaning and subtly different grammar.

The first is found in sentences like *Step the fuck down* or *Shut the hell up*. We'll call this the *Get-the-hell-out-of-here* construction, based on the earliest known attested use, from 1895, which was that, verbatim.[7] Superficially very similar but, as we'll see, clearly distinct is a second use of *the* + expletive. It looks like this: *That girl knocked the hell out of that piñata* or *I'm going to eat the fuck out of this lasagna*. Let's call this one the *Beat-the-devil-out-of-her* construction, after the earliest known usage from an 1885 romance novel: *Loubitza will beat the devil out of her when she gets her home*.[8]

Why should we think that *Shut the hell up* is a different beast from *Knock the hell out of that piñata*? After all, they have striking similarities. They seem to admit the same taboo words (*shit, hell, fuck*, and so on), and they display the same taboo form: *the* + expletive. But there are several reasons to think that you actually follow distinct rules for them: they have different properties and yield to different constraints.[*]

First, the expletive seems mandatory for the *Beat-the-devil-out-of-her* construction but optional for the *Get-the-hell-out-of-here* construction. We know this because taking *the fuck* or *the hell* out of *Get the hell out* or *Step the fuck down* or *Shut the fuck up* produces perfectly grammatical sentences: *Get out, Step down*, and *Shut up*. Their meaning just becomes a little less intense, as you would expect when you omit profanity. But the same doesn't hold when you take the expletive out of *The girl knocked the hell out of that piñata*. That produces the ungrammatical *The girl knocked out of that piñata*. Strange. Removing *the fuck* from *I'm going to eat the fuck out of this lasagna* yields *I'm going to eat out of this lasagna*, which can't be interpreted as describing the same sort of thing that *eating the fuck out of this lasagna* does. (It might be interpretable, but with a totally different meaning, where the lasagna becomes a container for the food being eaten.)

Optionality of the expletives is only one way in which *Beat-the-devil* and *Get-the-hell* differ. They also display different behavior when you apply

---

[*] Hoeksema, J., and Napoli, D. J. (2008) have a thorough and delightful exploration of these two constructions, which I've leaned on heavily for the following.

other grammatical rules to them. For instance, in English, you can make an active sentence like *John ate the carrots* passive, so that it becomes *The carrots were eaten by John.* When you try to apply the passivization rule to the two constructions we're looking at, you again see that they differ. Specifically, *Beat-the-devil-out-of-her* sentences can be made passive, but *Get-the-hell-out-of-here* sentences cannot. For instance, you can say *The piñata got the hell knocked out of it by the girl* and maybe even *The hell got knocked out of the piñata by the girl.* But take a *Get-the-hell-out-of-here* sentence like *Get the hell out of here,* and you'll be hard pressed to passivize it. Neither *The hell was gotten out of here* nor *Here was gotten the hell out of* does justice to the active original. These two types of sentences, although they superficially contain similar inserted expletives, actually behave quite differently. The best we can do is call *Get-the-hell-out-of-here* and *Beat-the-devil-out-of-her* different grammatical patterns and try to understand why they work the different ways they do.

And as I hinted at the outset of this section, we can find part of the explanation for their different and idiosyncratic properties in their histories. As its earliest use would suggest, the *Beat-the-devil-out-of-her* construction was originally patterned off of an existing sentence form in English. *They beat the devil out of her* is a sentence of the very same form as *They pulled the survivors out of the ship* or *They forced the mayor out of office.* Namely, there's a subject (*they*), a verb that describes acting on something with enough force for it to move (*beat, pull, force*), then an object that is forced to move (*the devil, the survivors, the mayor*), and finally a direction it is forced to move in (*out of him, out of the ship, out of office*). This pattern is often called the caused-motion construction because it describes someone acting on something forcefully to cause it to move.[9]

In other words, originally, sentences like *Loubitza will beat the devil out of her when she gets her home* were actually not unambiguous instances of a special *Beat-the-devil-out-of-her* construction. Instead, they were probably talking about actually acting on someone (*her*) via a forceful action (*beating*) to make something (*the devil*) move in some direction (*out of her*). Many examples of the *Beat-the-devil-out-of-her* construction remain ambiguous to this day, hovering between a caused-motion interpretation and a *Beat-the-devil-out-of-her* interpretation. For instance, does *If you keep misbehaving, I'll knock the hell out of you* mean that upon completion, the hell (namely the bad intentions and behavior) will have been removed from you? Does *They're going to beat the shit out of me* describe literal or metaphorical feces being punched out of the victim?

The origins of the *Beat-the-devil-out-of-her* construction explain its idiosyncratic behavior. The expletive is mandatory because *the hell* or *the fuck* or *the devil* was, at the origin (and perhaps continuing in some ambiguous cases to this day), an actual thing being acted on such that it would move. The caused-motion construction generally makes this particular component mandatory. For instance, you can't omit *the mayor* in *They forced the mayor out of office*; that produces the ungrammatical *They forced out of office*. And the *Beat-the-devil-out-of-her* construction, like the caused-motion construction it derives from, allows passivation without hesitation: *The mayor was forced out of office by them*. In other words, the *Beat-the-devil-out-of-her* construction behaves the way it does in terms of grammar because of patterns established when it was still being created.

The *Beat-the-devil-out-of-her* construction originated in ambiguity. But over the more than one hundred years of its use, this construction has expanded to include other cases that we can no longer interpret via the originating caused-motion construction. For instance, *I'm going to eat the fuck out of this lasagna* clearly doesn't imply that there's *fuck* in the lasagna that *I* will somehow remove via eating. And this construction seems to be on the move in that more and more verbs are being recruited to it. It now seems to many people perfectly grammatical to say *I'm going to sprint the fuck out of this marathon* or *After this semester, I'm going to know the hell out of physics*. These are clearly not about causing motion, but remarkably they still exhibit the hallmark grammatical properties carried forward from *Beat-the-devil*'s origins. The expletive is still mandatory: you have to say you'll *know the hell out of physics*; you can't say you'll *know out of physics*.* So our grammatical minds are littered with traces of the history that these particular profane expressions have traversed.

$ % !

Profanity has a grammar all its own, after a fashion, but it's not a tidy affair. If the grammar of a language is a system of regularities—uniform

---

* However, in some cases, it cannot now be passivized: *The hell will be known out of physics by me* and *Physics will be known the hell out of by me* are both quite ungrammatical. See Hoeksema, J., and Napoli, D. J. (2008) for an explanation.

patterns of behavior—then profanity simultaneously taps into some of the regularities and imposes a number of distinct subregularities all its own. Sometimes this produces utterances that, by any of the standard criteria, are not sentences at all. At other times it produces sentences that appear to be missing components or that flout the overriding rules of the language. Some of these subregularities we can explain by appealing to the meaning they're used to convey, the function they're put to, or the history of how they came to be.

When we broaden our scope a bit, it's fair to speculate that other types of language, aside from profanity, may work similarly. There are probably specialized subgrammars for each of the purposes we put language to. There's a special way we recite numbers and dates. Recipes have a particular formula (compare the standard English *Mix the eggs into the flour and beat them together* with Recipese: *Mix eggs into flour. Beat together*). So does speech directed at children or pets. And of course we know that different groups of people use different rules of grammar and that some people who can easily flit between such subgroups may even have the capacity for language in multiple distinct dialects of a language, with different rules of grammar. Although we began with profanity, a bigger question is really at play here. When we talk about grammar and try to understand how it works in the human mind, should we be talking about a single grammar for a language or a patchwork of subgrammars specialized for particular purposes? The structure of the grammar seems to be shaped by what you're trying to do with it.

I'm all for simplicity. And it would be far simpler if grammar would just keep to itself. Let words convey meaning so grammar can just be in the business of putting them together into bigger structures. But human language doesn't appear to work this way. Instead, specific grammatical choices seem to carry with them—or to be driven by—the meanings and functions they're paired with. The question that remains is, how much of language is like this? What's the balance between the specific, meaningful, idiosyncratic patterns we've been looking at here and the ostensible patterns of grammar that are truly general and truly meaning-free? Profanity raises the question, but we don't yet have the answer.

7

How *Cock* Lost Its Feathers

A thousand years ago, people living in the British Isles spoke various manifestations of Old English, the language that would evolve over the subsequent millennium into what we now know as Modern English. In terms of its words, what they mean, how they're pronounced, and how they're arranged grammatically, Old English is probably just as foreign to the contemporary speaker of English as is German or Dutch. For example, take the Old English version of the Lord's Prayer (to jog your memory, this is the one that now goes "Our Father who art in heaven, hallowed be thy name," and so on). In Old English, it looks like this, in its entirety:

> Fæder ure þu þe eart on heofonum; Si þin nama gehalgod to becume þin rice gewurþe ðin willa on eorðan swa swa on heofonum. urne gedægh-wamlican hlaf syle us todæg and forgyf us ure gyltas swa swa we forgyfað urum gyltendum and ne gelæd þu us on costnunge ac alys us of yfele soþlice.[1]

If you're lucky, you might be able to pick out a couple of vaguely recognizable words, especially once you know that a rough word-by-word translation into Modern English starts with "Father ours, you who are in heaven . . ." As you can see, *father* used to be *fæder*, and *heaven* was *heofonum*. Like great-grandparents depicted in a blurry black-and-white photograph, some of these words bear a family resemblance to their modern kin. Other words are totally unfamiliar. *Gedæghwamlican* means "daily." *Alys* means something like "redeem." The passage of time does this to a language. As the years tick by, words morph into progressively less recognizable forms, while others are replaced entirely.

The Lord's Prayer isn't the only preserved record from the ancient history of English—liturgical, scientific, legal, and literary texts enshrined in the museums and archives of the Anglophone world contain within them the necessary rudiments to trace out the histories of not just the sacred words of our language but the profane ones as well.

Illustrative of how profanity changes is *cock*, a word as old as English itself. In ancient records, it can be found spelled variously as *coc*, *cocc*, or *kok*—as you can see, orthography was just as inconsistent during the Middle Ages as it is during the Texting Age. But from its earliest recorded use in AD 890–897, we know that in the first millennium, the word referred to one particular kind of *cock*, the kind that crows.[2] Eleven hundred years later, the word has transformed. To contemporary American ears, of course, a *cock* is most likely something else, something not typically outfitted with a beak and feathers. The meaning of *cock* has changed radically.[*]

In the place of *cock*, contemporary American English speakers prefer to refer to the male of the species *Gallus gallus domesticus* using the word *rooster*. We can track this shift through the historical record of English. The Google NGram corpus provides a count of how frequently different words have been used over time, at least as recorded in the books that Google has scanned to date. The record gets less reliable the farther back you go, especially for relatively infrequent words, like *cock*, that only pop up now and then. But if you track *cock* over the past 150 years, you'll quickly see that it has come to be used progressively less and less over the years. The chart you see on the next page plots time on the *x*-axis—from 1850 to the present. And on the *y*-axis is the frequency of *cock* among all words in books from that particular year. You can see that *cock* has been quite infrequent for centuries: only about one word in 100,000 is *cock*. Presumably, that's because writers have other things to discuss than fowl and penises. But what's important is the change over time. By 2000, *cock* was used only about a third as often as it was one hundred years before. And lest you think this is merely because male chickens have progressively fallen out of favor as a writing topic, compare the downward trend of the *cock* curve

---

[*] If Mark Twain had wanted to write *A Connecticut Yankee in King Arthur's Court* as a farce, he could have gotten a lot of mileage out of this. Think about the riotous malentendus when an uptight courtesan proclaims *I'm hungry for a cock tonight* or *I woke up this morning to that damned cock again.*

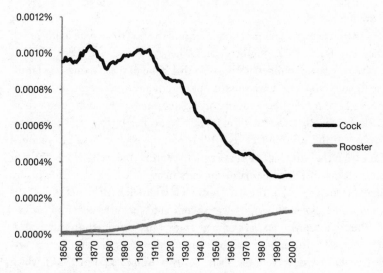

The decline of *cock* and the rise of *rooster*.

with the upward trajectory of *rooster*, below it. *Rooster* was basically un-attested in 1860 and has since risen to a nearly *cock*-like level. People are using *cock* less and *rooster* more.

Wait a tick. Does the Google corpus really contain no recorded in-stances of the word *rooster* in 1860? Or in 1850? Or anywhere in between? Zero? How could this be? *Rooster* feels like a word as old as the language. Isn't it?

We'll get to that.

But first, to sum up what we know so far: *cock* was once an innocuous animal term. Now, however, with the twenty-first century in full swing, if you ask a rancher about his livestock, at least in the United States, he's unlikely to say that he has three hundred pigs and five hundred cocks. The original sense of *cock* has shriveled up and been replaced by *rooster*, a word that seemingly appeared in the nineteenth century.

What's special about profanity—what makes it distinct from other types of language and particularly important to study—is that all of these facts repeat themselves again and again in one profane word after another. When profanity evolves, it tends to follow certain recurring patterns that open a window onto how and why languages change.

$ % !

Exhibit B is *dick*.

There are many famous *Dick*s. Here are the first that come to my mind: Dick Van Dyke, Dick Smothers, Dick Cheney, Dick Cavett, Dick Clark. If you're twenty or younger, it's possible that none of these names are familiar to you—you who have missed out on unique American cultural treasures like *Mary Poppins* and *American Bandstand*. If you don't know these Dicks, it's because they're all old. Their respective birth years? 1925, 1938, 1941, 1936, and 1929. In fact, as it turns out, there are a lot of guys named *Dick* from the early half of the twentieth century and well before. The first attested use of *Dick* as a name appears in 1553,[3] just eleven years before Shakespeare was born. The Bard himself might well have had *Dick*s as contemporaries. For a while, *Dick* was as common a name as *Tom* and *Harry*.

But you know who isn't named *Dick*? Most anyone born after 1968. It's impossible to get hard numbers on nicknames, ephemeral as they are. But we do have data from the Social Security Administration (SSA), which tracks the names given to babies born each year.[4] As you can see from the graph on the next page, *Dick* was once a quite popular name to give to young boys. To be clear, these aren't babies named *Richard* and nicknamed *Dick*. Oh no. These are babies whose birth certificate proudly displays their given first name as *Dick*. According to the SSA, there were eight hundred of them per year in the 1920s and 1930s. As you can see, *Dick* started petering out after reaching its high point in the early 1930s. Newborn babies were still being named *Dick* through the 1950s, but by the 1960s, *Dick* was clearly on its way out. And then, starting in 1970, no more *Dick*s. Instead, *Rick*s.

The similarity between *cock* and *dick* is striking. In both cases, at one time, the word is entirely anodyne. At a later point, it means something different and profane, and the role it originally played passes to a different word altogether—*rooster* or *Rick*. The list of profane words in English is largely a list of words with similarly humble, inoffensive origins. *Bitch* used to refer just to female dogs. A *faggot* used to be a bundle of sticks. *Ass* used to be a donkey. And so on.

For a language scientist, a trend like this screams out for a deeper look. How are these particular words selected to become profane? Where do the old meanings go? And where do the replacement words come from—words like *rooster* and *Rick*? So what follows is an outline of the career arc that *cock* and *dick* and many other taboo words have scratched out, from banality to profanity and eventually to obscurity.

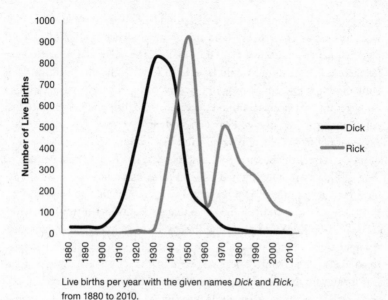

Live births per year with the given names *Dick* and *Rick*, from 1880 to 2010.

Step one: a word extends its meaning

Before *cock* and *dick* became profane, they already existed as words but with different meanings. This is true of most profanity. I mentioned *bitch*, *ass*, and *faggot* earlier, but other examples abound. *Jesus* used to just be a nice name for a Jewish boy, for example.

The history of *fuck* appears similar, despite fanciful stories you may have heard to the contrary. There's a claim floating around that *fuck* was created as an acronym—perhaps for something like *For Unlawful Carnal Knowledge* or *Fornicate Under Command of the King*. Acronyms are certainly a bountiful source for profanity—as demonstrated by recent examples like *MILF* and even *GILF*, *THOT*, and *WTF*. But the historical record leaves no reason to believe that *fuck* was born as an acronym. A key bit of evidence is that *fuck* has apparent cognate words in languages related to English. As far as we can tell, *fuck* is related to its German equivalent *ficken*, which is roughly synonymous with its English cousin (though less widely used and less profane), as well as Dutch *fokken* ("to breed") and Icelandic *fjúka* ("to be tossed by wind").[5] This suggests that the word's ancestor can be traced back thousands of years to a time before the languages that evolved into modern-day German and English diverged. So if *fuck* were formed as an acronym (and there's no evidence it was), this would have happened thousands of years ago, and the particular words whose first letters it spelled out would have been totally different from words of contemporary English. Words like *carnal, knowledge, command*, and so on were not part of the common ancestor of English and German and so could not have been used to form an acronym.

So if not from an acronym, how did *fuck* become the go-to swearword of modern English? Because *fuck* is so old and has had its meaning for so long, it's hard to know if it had a life prior to its profane one—we simply have very little in the way of preserved written texts that go back that far. But there's a bit of indirect evidence that in its ancient history it derives from an Indo-European root from thousands of years ago meaning something like "to strike," "to stab," or "to stuff."[6] That is, before it came to refer to copulation and before it became profane, it was likely a mundane Indo-European verb that described a simple and inoffensive physical action.

Precisely when it extended its meaning to specify a particularly lurid type of striking, stabbing, or stuffing is currently unknown, though this appears to have happened at least as long ago as the fourteenth century. Medievalist Paul Booth recently uncovered the earliest known record of the word to date, in legal documents from 1310 identifying a man as Roger Fuckebythenavel; parsed out, that makes *Fucke by the navel*. Booth explains that the name "could either mean an actual attempt at copulation by an inexperienced youth, later reported by a rejected girlfriend, or an equivalent of the word 'dimwit,' i.e., a man who might think that that was the correct way to go about it."[7] This suggests the word has been doing dirty work for at least seven hundred years.

This common pathway to profanity suggests that although the Holy, Fucking, Shit, Nigger Principle tells us which semantic fields profane words are most likely to be drawn from, those words often have an even earlier history. Before they refer to genitalia or copulation, they have other lives, as words referring to farm animals or commonplace actions like hitting, for instance. So the first step toward profanity is actually to acquire a Holy-, Fucking-, Shit-, or Nigger-related meaning.

The ways they add these new meanings are largely typical for words. A common mechanism is metaphor: words commonly come to be used for something perceived to be similar to the original meaning. Consider why it is that we refer to the face of not just a person or animal but also of a clock.

Although a clock face has no eyes, nose, or mouth, one may apprehend global visual resemblance to a human face. This explains why the word, first recorded referring to part of a human in AD 1300, was extended to clocks as well soon thereafter.[8] In this case, the similarity between the two faces is superficial and visual. *Cock* is probably an example of this as well. As early as around AD 1400, we have records of *cock* referring not just to roosters but also to the male member. And the reason may well be superficial visual similarity between a rooster and a penis. Similarly, the name *Dick* gained an additional meaning in the 1870s. It's not the one you're thinking of. It came first to refer to a riding whip (how, we don't know). From there, in the following decades, *Dick* was extended once again to a new meaning, this time referring to the male member (according to some sources, perhaps initially in military usage).[9] The superficial visual similarity between the handle of a whip and the male member is probably responsible—though, importantly, it's not sufficient. Many words denoting things longer than they are wide have not over the years come to refer to the male member. So chance is surely at play in this part of the process.

But metaphor doesn't restrict itself to the surface. Words can also find themselves metaphorically extended to new meanings due to deeper, structural connections. We now use *face* in expressions like *on the face of it*, which can refer to things that don't have any physical manifestation at all: *On the face of it, this theory doesn't have a leg to stand on!* This kind of deeper metaphor has also generated some of our profane words. Consider *bitch*. Assuredly *bitch* was not extended from dogs to people due to superficial visual similarity. Although it's now used to refer to a malicious or unpleasant person, particularly but not only a female one, its original foray into humanity was to denote a lewd or sensual woman. Again, there's little visual similarity between a female dog and such a person. But extending *bitch* in this way might have constituted an appeal to perceive an abstract similarity between the behaviors of female dogs, particularly during estrus, and lascivious women.[10]

Metaphor might explain how the plausibly anodyne ancestor of *fuck* came to refer to sexual intercourse. Presumably I don't need to explain the superficial visual similarities between stabbing, stuffing, or striking and certain aspects of intercourse. These ways that meanings get extended are in no way unique to profanity. What's special about profane words like *dick*, *cock*, *bitch*, and *fuck* is that the meanings they gained were in an optimal position to become profane. They referred to copulation, genitalia, and so on.

Step two: dissemination

In order for a word with a new meaning to become profane, that change has
to catch on. Let's say that someone introduces a change into a language—
for instance, someone starts using *cock* to refer to the male member. In
order for it to have the string of consequences we now know that change
had, other people have to start using it too. The thing is that most changes
don't catch on. For instance, consider a word that I myself invented, *hummerbird*. This word is meant to refer to someone who flits from sexual partner to sexual partner performing oral sex.* I hope you find some use for the
word. But if you do, you'll be basically on your own, because it has most
certainly not caught on. No matter how useful you and I find *hummerbird*,
no matter how frequently we use it, unless other people start using it too,
the change will be lost as soon as you and I stop. You might know people
who make up new words and then try to get other people to use them.
Those people usually fail—almost all changes introduced into a language
die off before they ever catch on. Urbandictionary.com is a graveyard for
words people thought up that were going to be their big claim to fame,
only to be entombed in the obscurity of two upvotes and three downvotes
forevermore.

Occasionally, though, a change gains traction. Someone said the word
taint, and the English-speaking world was forever changed. Why? It's
worth trying to understand what makes a change more or less likely to
diffuse through a community. Do new words contain intrinsic properties
that make people want to reuse them? And how do words spread? Who has
to use them for other people to decide they want to use them too? How, for
example, did *cock* manage to spread its wings after losing its feathers?

We know a little about how this works. The success of a new word or a
new use for an old word depends on at least three things. First, the intrinsic
properties of the word itself matter.[11] All innovations are not equal. Some
words are shorter, easier to pronounce, or easier to remember than others,
which may contribute to their eventual success. With English profanity in
particular, as we've seen, there's also a sound pattern (one syllable, with a
consonant at the end) that makes words sound like swearwords. So a word

---

* It has come to my attention that certain birders, particularly in Texas, already use the word
  hummer bird to refer to hummingbirds. Influential though the birding community may be, I
  don't believe it's responsible for obstructing my innovation from catching on.

with this particular phonological property is more likely to succeed, all things being equal, than one that doesn't.

Another intrinsic aspect that appears important is how transparent (versus opaque) the word is. There's a sweet spot for potentially profane innovations. A word is transparent if you can easily figure out what it means from how it sounds; conversely, opaque words are inscrutable. Opaque words are hard to remember and as a consequence may be less likely to spread. So if you started calling the perineum the *baint*, which is completely opaque, you probably wouldn't find adolescents using the word in five years. If you called it something totally transparent, like *interorgan region*, it wouldn't likely make much of a splash. But there's a middle ground between transparency and opacity. Words can be "motivated." For example, *taint* has been quite successful. And that might be due to the fact that it lives in the Goldilocks region of motivatedness. It's not obvious to an outside observer why the perineum would be called the *taint*, even if *taint* does have another negative meaning (a stain or black mark). But if you're an insider who knows the perhaps apocryphal origin story,* then it takes on a whole new life. It's motivated.

A word that's totally opaque faces a long, uphill climb to acceptance. At the same time, a word that's completely transparent may fail to bestow cachet on people who know it. But *taint* is right in the middle of these two extremes. The same is true of *MILF* or *butterface* and plausibly, at their origins, *bitch, cock, fuck*, and others. Motivated words may be more likely to catch on than transparent ones because they require a little additional knowledge to interpret. They're like a secret code that only insiders have the key to cracking. Opaque words are equally novel to everyone, while motivated words are, in a way, transparent to insiders only.

And that leads us to the second reason word changes may spread. Often changes in language catch on precisely because of the social functions they serve—in the case of profanity, allowing people to feel and identify themselves as part of specific social in-groups. Said another way, if a teenager's parents could figure out right away what the child meant by *hummerbird*, it would take the fun out of it. So she might not use it. In other words, just as

---

* Ostensibly, the *taint* is so named because "Taint your ass, taint your balls." I can't verify this account—it's equally plausible that this is a post hoc folk etymology and that the original motivation for the word really comes from the other meaning of *taint* ("uncleanness"). Either way, Goldilocks should be happy.

life forms thrive when they find a suitable environmental niche to exploit, so words thrive in fertile linguistic niches.

For any new word or new way of using an old word, we can ask, what niche does this innovation satisfy? Let's look at an example of a successful innovation, the *tear someone a new one* pattern from the last chapter. This relatively new invention describes the very frequent and frequently talked about situation in which one person verbally or physically attacks another. Interpersonal conflict is one of our favorite things to talk about. So there's demand for language to describe it. Now, it's true that there are other ways to talk about conflict. For instance, focusing just on verbal conflict, you can *chew someone out, bite someone's head off, give someone an earful*, and so on. But *tear someone a new one* expresses this meaning in a courser, more vulgar way. Other successful recent innovations like *MILF* and *THOT* also claim semantic territory that was previously underpopulated.

And finally, the third factor in a word's success is in no small measure who is using it—the status of its users and how they're connected within the network of people who make up a language community. Even assuming that you have a good innovation, which satisfies a particular linguistic niche and is intrinsically promising, that doesn't guarantee it will spread throughout a population. Changes disperse through a language community in somewhat predictable ways. They spread first among people who talk to each other most and especially those who identify as belonging to the same social groups. We can quantify this in studies of how linguistic innovations spread over social media. For instance, one recent paper looked at what predicted whether a new word used by people in one city would spread to other cities.[12] The researchers found that cities closer to each other are more likely to share new words and, in addition, that similar socioeconomic and ethnic makeups increase the likelihood that a change will spread between populations. So both geography and demographic similarity are important. We can interpret this to mean that changes will spread among people who communicate with, and think they're similar to, each other.

But change doesn't flow in every direction equally. In any group—kids hanging around a locker room, coworkers hanging around the break room, or gamers hanging around an online chat room—individuals will be unequally dominant or influential. If the more influential ones use language in a new way, then others are likely to follow suit. Of course, the changes don't spread as fast in the other direction. Differential influence has the most outsized effects in the mass media. Stephen Colbert can coin a term (and does frequently), then use it on his television show, and it will catch on, whereas one of his millions of fans will never enjoy the same success, no matter how hard he or she yells at the screen. Many profane words gain traction through the media.

But this doesn't necessarily mean that's always where those words originated. For instance, many people believe the word *MILF* to be the brainchild of the writers of the 1999 movie *American Pie*, in which the acronym is spelled out amid boisterous chanting by teenage boys. But a little Internet sleuthing reveals that it dates to at least a few years earlier. The earliest attested use I was able to find dates from 1995, in a usenet post about a *Playboy* spread. Here's the usenet post I found* (though this might not be the earliest use—T. J. Kelleher reports an instance several years earlier in a slang dictionary by University of California, Los Angeles, linguistics students[13]):

> WOW! I saw the pictorial in the Feb issue and boy was I impressed.
> Those moms are babes!! Almost unbelieveable [*sic*], especially that union worker one towards the front, you almost have to look twice . . .
> We have a term for it around here, its [*sic*] called "MILF"
> It stands for "Mothers I'd Like to Fuck."
> Maybe that is what they should have titled the section   :)
> -Just my $0.02
> Mike
>
> ===
> Michael Andreano   Chi Phi Fraternity
> Hoboken, NJ   Stevens Insititute [*sic*] of Technology

Although the term had a life prior to the movie, its current popularity no doubt results from its use on screen rather than on a *Playboy* fan newsgroup. Likewise, the popularity of the word *Johnson* for "penis" exploded after it appeared in the 1998 film *The Big Lebowski*. But just as *MILF* has

---

* I contacted Mr. Andreano to ask where he thinks the word originated but didn't hear back from him. In case you're concerned that including his name here is a violation of his privacy, I was too, and I originally was going to anonymize the post. But I changed my mind for two reasons. First, the post is accessible online; a search that includes even a little of the text of the message will reveal the author's signature. And second, I thought it would be a useful reminder that everything you do online will remain publicly accessible in perpetuity.

roots preceding the release of the movie, so the Nihilists in *The Big Leb-owski* were far from the first people to talk about harm to people's *Johnsons*. Take a look at the following entry from Walter Butler Cheadle's *Journal of a Trip Across Canada*, which predates the 1998 Coen Brothers film and de-scribes an expedition from Quebec across the Canadian Rockies to British Columbia: "Neck frozen. Face ditto; tights ditto; Johnson ditto, & sphinc-ter vesicae partially paralyzed." Walter Cheadle put pen to frigid paper to write that journal entry in 1863.[14]

In modern times, it's comparatively easier to track how words dissemin-ate throughout a speech community. For instance, we know that *ctfu* ("cracking the fuck up") spread mostly from Cleveland to a number of other mid-Atlantic cities, as you can see in the figure on the next page.[15] And we know this because people leave quantifiable records of their lan-guage use in the form of GPS-coded tweets.

But we have no such luxury for changes that occurred in the deep his-tory of English—pre-Internet. So we know little about exactly how *cock*'s new meaning spread throughout the English-speaking world starting in the fifteenth century. But we do know what niche it filled. Every language has a way to describe human sexual organs. They're pretty important, cul-turally, biologically, personally. It seems reasonable to assume that the new use of *cock* was somewhat motivated—there's a passing similarity between a rooster and a penis—and we now know that as a closed monosyllable, it conforms to the sound pattern of English taboo words. We don't know, however, who used it first or how it spread. It's very likely that it diffused through networks of people talking to one another, where the status of the people using the word influenced its ultimate success. And mass media of the day—songs and poems and eventually books and newspapers, for in-stance—may have played a role, as they do to this day in spreading changes in a language.

Step three: all the action is in the reaction

So we've arrived at the point in our story where a previously unremarkable word has gained a new meaning that has spread within a language commu-nity. *Dick* now means not only *Richard* but also *penis*, and a lot of people are using it this way. But to be clear, these words have not yet become profane. *Dick* and *cock* came to refer to the penis in the fourteenth and nineteenth centuries, respectively, but those shifts by themselves didn't make the words

Weeks 1–50

Weeks 51–100

Weeks 101–150

Tracking the use of *ctfu* over time via GPS-coded
tweets. Source: J. Eisenstein et al. (2014).

profane. *Fuck* has referred to sexual intercourse since at least the fourteenth
century, but it only became so taboo as to disappear from the correspon-
dence of upper-class ladies at the end of the eighteenth century.[16]

The same is true of the word *cunt*. It had been in general use for centuries with the meaning "vagina" before people began to find it offensive.
In fact, it was apparently of such widespread and untainted use in early
English that it even appeared in people's names. The *Oxford English Dictionary* lists several names built from *cunt*, like John Fillecunt (1246), Robert

Clevecunt (1302), and Bele Wydecunthe (1328). *Cunt* also appeared in place-names. It was a long-standing tradition in England to name streets after the economic activities that predominated on them. This produced names like Silver Street and Fish Street. And in many English towns of the thirteenth through the sixteenth centuries, when prostitution was concentrated in specific places, these streets were commonly named some variant of *Grope-cunt Lane*. Let's be clear: that's *Gropecunt*, as in *grope* plus *cunt*.[17] Not very subtle, England. And *cunt* was so pervasive because, at the time, the word was a straightforward and inoffensive description of female genitalia. It even shows up in Middle English medical texts like Lanfrank's *Science of Cirurgie*[18] from AD 1400:

> *In wymmen þe necke of þe bladdre is schort, & is maad fast to the cunte.*
> "In women, the neck of the bladder is short, and is made fast to the cunt."

This shows that becoming profane is a social change, not a seman-tic one. Like other human behaviors, language use is at the whim of cul-tural beliefs, norms, expectations, and prohibitions. Consider the way that norms for acceptable attire change over time and from culture to culture. We often think of history as making inexorable progress toward increased freedom, and you can take, for instance, norms about how people dress at the beach as an example of this, from Victorian full-body tunics through the bikini and speedo a hundred years later. But clothing norms ebb and flow. Over the last several decades, Afghanistan and Iran, for example, have seen substantial decreases in the parts of the body—particularly the female body—that people can acceptably show in public. The same goes for other behaviors, which are accepted or prohibited differently over time and across cultures. For example, spitting on the ground is illegal in Hong Kong, where it carries a fine of thousands of dollars, while Taiwanese people consider spitting out bones onto the table or floor while eating pref-erable to using their hands.

The point is that while beliefs about what behaviors are permissible in what contexts may often have some moral, spiritual, religious, scientific, or medical foundation, they're also culturally relative. What's sanctioned in one place, or at one time, or when done by one person might be socially unacceptable in other circumstances. And so it is with language.

There are different reasons why people at a given time or place might find a word unacceptable. Perhaps they find it offensive because it leads to thoughts that they don't wish to have—about taboo concepts, for instance. Or perhaps they find the word insulting to them or a social group they belong to. (We'll look in more detail at slurs and where they come from in Chapter 10.) These personal feelings about acceptable behavior become social norms when enough people engage in actions that constrain the behavior. Everything from public indecency laws down to muttered indications of displeasure enforce norms about clothing or spitting: *Can you believe he's wearing THAT?* And these same types of enforcement take linguistic innovations—new words or new ways of using words—into the realm of profanity. Marginalizing a word through direct interpersonal action, like ostracizing or punishing people who use it, or indirectly through social and state actors that impose censorship will render those words profane.

Step four: the balance shifts

Once a word has gained additional meaning and once that additional meaning has come to be socially proscribed, as happened for *dick* and *cock*, the new meaning starts to color any use of the word. When words have multiple meanings, as many words do, it's often impossible to use them in one way without activating other possible interpretations in people's heads. For example, when you hear a news anchor tie up a report featuring a variety of exotic animals, including a lion, by saying, "Nice pussy!"[19] it's simply impossible to contain your interpretation to the single, intended sense of *pussy*. You shouldn't feel guilty about this. It's just how your brain works. When a word has multiple meanings, you systematically and automatically activate the different meanings, regardless of whether the speaker intended them or not.[20]

This is equally true for *cock* and *dick*. At the dawn of the twentieth century, *Dick* was living parallel lives. At the same time that it was referring to sexual organs, there were still children, adolescents, and adults named *Dick*. The word had several meanings. That led to confusion, some unintentional, some quite deliberate. For an example of what I can only imagine is the latter, enjoy the actual campaign button for Richard Nixon reproduced on the next page.

This effect—the spontaneous activation of the multiple meanings of a word—is so overwhelming that only under the rarest conspiracy of conditions does *pussy* evoke only the thought of a cat.[21] For that to happen, the context has to be strongly compatible with the intended sense but totally

Source: Gene Dillman of Old Politicals Auctions, www.oldpoliticals.com.

incompatible with the unintended one. *What a nice pussy* is compatible with both interpretations, and you really have to stretch your imagination to find contexts only consistent with the feline one. Perhaps *I predict that my cat is going to give birth to a calico pussy* would be an example. But maybe not. And second, in order for the unintended sense to remain out of mind, it has to be a peripheral and infrequent sense of the word. The genitalia sense of *pussy* does not currently qualify as peripheral and infrequent. *Tea-bag* might be a better example of a word that, at least for many people, has an infrequent and peripheral profane sense. But profane senses for words can become central quickly. As we saw earlier, profanity grows deep roots into the brain's emotion systems. Merely seeing or hearing a word that has a profane meaning activates this meaning immediately, context be damned. There's no number of cat videos that would keep *pussy* from activating the other meaning in your mind.

And people seem to know this, whether consciously or unconsciously. Double entendres make use of this very feature of human language processing. Writers play on it all the time. The James Bond character Pussy Galore was not thus named blithely. On *The Simpsons*, in an early episode

("Treehouse of Horrors III"), Marge is about to board a ship, at which point Smithers (Mr. Burns's assistant of ambiguous sexuality) comments, "I think that women and seamen don't mix." Mr. Burns's response: "We know what you think."*

The safest strategy for the speaker who doesn't want to make a verbal misstep is simply to avoid those words in general that have a possible second meaning. I know that I for one simply don't ever use the word *pussy* to refer to a cat, period. I don't refer to a rooster as a *cock*. I will call my dog a *bitch*, but that's always with malice aforethought, and I only do it when I think the audience will be forgiving or when I want to be a little edgy.

So the consequence is this: When words acquire new meanings through natural processes of meaning extension, and when those new meanings are profane, then to the extent that speakers feel that those profane meanings will have negative consequences, they'll start to use the words only when the profane meaning is what they intend. And they'll avoid using the word with the original meaning. In the case of *Dick*, we can see why some people might find the ambiguity untenable. By the 1960s, it came to a head. The profane use was too much to overcome, and people gave up on *Dick* as a name. Similarly, we can presume that something similar happened with *cock*—that at some point, the profane meaning overwhelmed the original galline one.

And just like that, the balance can shift, often quite quickly, from an older meaning to a new one. This is something special—or at least particularly pronounced—about profanity. We don't readily see it with nonprofane word change. For instance, although *cell* now refers not just to a jail room and the basic structural unit of biological organisms but also to the thing in your pocket that you watch cat videos on, this hasn't led us to stop using the word *cell* in its old senses as well. Same with *mobile*, which used to (and still does) describe the thing hanging over a crib, even though it now also refers to the cat video machine in your pocket (although some people might pronounce the two *mobiles* differently). That's not to say that new meanings can't usurp old ones unless they're profane. Certainly they can. But they're not as likely to do so, and they don't do so as quickly or as completely.

To be clear, this shift doesn't have to happen. Although the earlier senses of *cock* and *dick* have largely fallen out of favor in American English, most other varieties of English have exhibited no such shift. *Cock* is still in

---

* The astute linguist will have noticed that this is actually an example of homonymy—*seamen* and *semen* are distinct words that happen to be pronounced similarly. Nonetheless, taboo homonyms, as you can tell, exert the same tug on the mind that words with multiple senses do.

favor as the default term for the rooster in most varieties of English spoken in Great Britain, for example. Clearly, in some times and places, speakers of particular language varieties are more comfortable with unintended innuendo. But the situation is unstable.

Step five: the replacements

So we've seen how, once a word gains a new, profane meaning, this new meaning can start to push out the old one. But this leaves a logical gap in the language. You need a way to refer to roosters, and you need a nickname for Richards. What do you do?

Sometimes, there's already a good alternative out there. As we saw, when *Dick* grew a new meaning, *Rick* stepped up and took its place. (You can see this in the SSA naming data chart. Ever since 1970, there have been more *Ricks* than *Dicks*.) Similarly, as *broad* came to have a derogatory connotation for a woman, it was replaced in the name of the track-and-field event that became, in the late 1960s, the *long jump*. And although it's a little cumbersome, *female dog* has largely (but not entirely) replaced *bitch*.

In other cases, there doesn't already exist a viable alternative to a word that has been made unusable in polite contexts. In this case, you have no other recourse than to make up a new word. People make up new words all the time in certain regular ways.

Although *rooster* seems like a word that should be as old as English, as we've seen, it's a relatively recent addition. The first recorded use dates from 1772, in the diary of a Boston schoolgirl.[22] The word was actually manufactured by design. It was the end of the eighteenth century, and the balance of *cock* had shifted to the point where its "male member" meaning had gained prominence. As a result, to simply talk about a rooster, puritanical Americans had to tread delicately around this second taboo meaning of *cock*. So they simply invented a new word, *rooster*. And the way they did it was pretty straightforward. They already had the word *roost* at their disposal, which, as it still does, described the place where chickens hang out. There was also a verb *to roost*, which denoted the action of hanging out in that place. And these linguistic innovators figured that just adding *-er* to the end, in the normal way that English is fond of, would create a new word that described something that roosts without the "penis" meaning attached. And they were right. *Rooster* has been largely free of profane connotations for the two hundred or so years it's been around.

This rule allowing us to add -*er* to a verb to create a new noun is particularly robust in English and pervades both profane and nonprofane language. Other mundane examples are *catcher* and *pencil sharpener*, and profane ones include *cock-blocker*, *shit-eater*, and *muff-diver*. But the application of a rule to generate a new word is only part of the story. Subsequently, *rooster* and all these other constructed words have to come to be used with this specific new meaning. They have to become "conventionalized"—the linguistic community has to decide that this word will be preferentially interpreted as having this specific meaning. We now agree, by convention, that *rooster* refers just to the adult male chicken and not to anything that happens to be on a roost. For instance, if an egg or a nest were on a roost, we wouldn't refer to it as a *rooster*. It's the same with other words invented using normal English processes. *Cock-blocker* has become a conventionalized word in English. We can tell because it has a specific meaning—it only refers to one type of *cock* (the type without feathers) and only to blocking *of* (and not *by*) that *cock*. *Motherfucker* is another word created using the same noun-verb-er template in surprisingly recent history; it is first recorded in the twentieth century![23] But since then, it has come into very frequent use and now merits being considered a word in its own right. And like *rooster* and *cock-blocker*, *motherfucker* has a specific, conventional set of meanings.

Adding a suffix to an existing word, as in these cases, isn't the only way to coin new words, of course. As we've already seen, totally new words can sprout in a language, assembled from the generative grammatical and lexical resources that the language makes available. Some are constructed by blending together existing words—merging them together to form a new, so-called portmanteau word that sounds something like a combination of them. For instance, *mangina* is a blend of *man* and *vagina*. In the same way, *fucking* and *ugly* give us *fugly*, and *pornography* plus *cornucopia* yield *pornucopia*.

Any of these tools for innovating words can create replacements for words tainted by new, profane meanings and can fill a gap in the language.

So, to take stock of where we are, we've now seen that banal words can change meaning to the point where they are candidates for becoming taboo. In the eventuality that this happens, people start to avoid using the words in their earlier senses and need to find or create a replacement, which the language in general provides various tools for. One small change in the meaning of one word can have downstream consequences for other words as well.

Step six: minced oaths

There's another way around using profane words—another externality of words becoming taboo. Once people start avoiding a word in certain circumstances, what do they say instead? Let's say you're in pain because you slammed your finger in a car door. Or you're angry because your neighbor is laughing at you for slamming your finger in the car door. This is prime profanity territory. And yet, if for reasons personal, religious, cultural, or otherwise, you decide that an expletive *fuck!* or *shit!* is out of line, then what do you do?

One option is to say something that sounds like a particular taboo word but isn't, like *frick* or *dang*. These are known as "minced oaths." Sometimes these minced oaths are actual, existing words used in new ways, like *shoot* or *fudge*. They get more elaborate, too, like *cheese and rice!* as a minced oath for *Jesus Christ!* or *Shut the front door!* to replace *Shut the fuck up!* But minced oaths can also be entirely new words. For example, over the years, people have come up with novel ways to avoid saying *Jesus Christ*, like *sheesh*, *gee whiz*, or *crickey*. Other invented minced oaths like *gosh*, *shucks*, and *frig* sound equally similar to the words they're used instead of.

There's a fine line to walk with minced oaths. They have to sound similar enough to their taboo target to be recognized as standing in for it. But they have to be different enough for the speaker to maintain plausible deniability—no, of course I didn't mean that filthy word (wink, wink). Quite often, the sounds at the front end of a minced oath are similar to the taboo word they're shadowing, as in the examples above. This might be related to the strategy that some people with coprolalia deploy—reshaping an expletive once it's started.

But minced oaths can also rhyme with the profane words they're replacing, and some of the most elaborate minced oaths to be found occur in a British form of slang called Cockney rhyming slang.[24] This linguistic art form starts by taking an expression that rhymes with a swearword. For instance, *Richard the Third* rhymes with *turd*. *Cattle truck* rhymes with *fuck*. So that expression comes to stand as a sort of minced oath for the profane word. But rhyming slang takes it one step further and uses some other, nonrhyming word from the same expression as a disguised cue for the whole thing. So instead of saying *cattle truck* or just *truck* to mean *fuck*, people say *cattle*. To mean *fuck*. Got it? OK, England, that's a little subtler.

A more concise strategy is to abbreviate the profane word. As mentioned in Chapter 1, the now rather tame (but formerly very profane) *zounds* is an abbreviation of *God's wounds*. In the most extreme case, the offending word is reduced to a single letter, as in acronyms. *WTF* or *MILF* take the first letters of the words in taboo expressions. Other similar cases are *BJ* and *T&A*. And for the most potent words, we can even refer to them simply by one letter. The *F*-word isn't the only word in English that starts with *F*—it's not even close to being the most frequent one. (That crown belongs to *for*, which is among the top twenty most frequent words in English.)[25] Nor are the *C*-word or the *N*-word particularly outstanding representatives of the language, except in terms of their offensiveness. These acronyms can be used to refer to the words in question (*He said the N-word!*), but they can also be used as replacements for the words themselves in normal use. For instance, *F* replaces *fuck* in expressions like *F off!* or *F you in the A!* or, of course, *What the F?*

Minced oaths and abbreviations have the advantage of preserving enough of the sound or spelling of the unsaid term that the sufficiently perspicacious audience can easily retrieve the intended but unsaid word. But some situations warrant a more delicate approach. If you don't want to be on the hook for even having thought about the profane word that you subsequently didn't say, then you need a euphemism. For example, people often want to avoid talking about death directly using the word *die*. Instead, they retreat to words and expressions that refer to the concept indirectly, such as *to pass on* or *to leave us*. Similarly, talking about the male member as a *Johnson* or a *wiener* is a way to euphemistically avoid overtly vulgar language.

As a case in point, below you can see a whole stable of words and expressions that all refer to broadly the same thing but with varying degrees of likelihood of offending sensibilities. The words and expressions toward the top, like *powder my nose* and *visit the ladies' room* are euphemisms. They tend to incorporate words that do not refer directly to the described act but instead mention related, innocuous activities.

powder my nose
visit the ladies' room
go to the bathroom
use the toilet
defecate
take a shit

cop a squat
do some paperwork
drop the kids off at the pool
pinch off a loaf

As you work your way down to the bottom of the list, you find ways to describe the same act much more vividly. These terms at the bottom go as far toward impropriety as the euphemisms at the top of the list go away from it. They're filling another niche. Whereas the euphemisms at the top fulfill the desire to describe a taboo topic in the most linguistically hygienic way possible, the dysphemisms at the bottom satisfy an urge to offend, impress, or entertain through lurid, evocative language. Oftentimes they describe vivid details of the described event (like *squat* or *pinch*). Sometimes they metaphorically describe the activity in terms of something else that itself evokes vivid imagery (like *loaf*). This makes for language that's particularly graphic, creative, or descriptive, while crucially not being strictly profane. A child who says she needs to *pinch off a loaf* might be chastised for being disgusting but not due to any of the specific words she selected. Dysphemisms abound, especially around taboo topics. In place of *to die*, people might choose to use stronger words like *to croak, to bite it, to eat it*. And for our current purposes, the most important bit is that in all of these cases, the meanings of existing words have been extended to cover new ground—*croak* now means more than just "to make a croaking sound."

Dysphemisms are just the other side of the same coin that euphemisms and minced oaths are embossed on. Profane words have external effects on the rest of the language, as people scramble to not say, or say without saying, or say even better, words that they know are taboo.

Step seven: change is the only constant

Words continue to change throughout their lifespans—becoming profane isn't an end point but a waypoint. This is clearly visible in the histories of *dick* and *cock*. For instance, even after gaining anatomical meanings, they came also to refer to an unpleasant person. And although that person originally had to be male, recent usage shows that for some English speakers, maleness is not a prerequisite to being labeled with one of these words; nor does it necessarily shield one from being addressed as *bitch* or *cunt*.

Even more radical changes are apparent in the history of profanity. Some words change their grammatical category, or part of speech. Of course, we've already seen how *dick* and many other profane words have become vulgar minimizers in expressions like *You don't know dick*. But that's just the beginning. Although *dick* and *cock* started their lives as nouns— referring to things—they soon came to be used as verbs as well. *Cock* gained its verb use—*to cock*—in the seventeenth century, and *dick* was verbed, as in *to dick around* or *to dick someone over*, only in the twentieth century. And other examples abound. For instance, *tea-bag* now acts as a verb as well as a noun, as in *Dave fell asleep at the party so Ray tea-bagged him*. (That means that Ray put his scrotum on Dave—and knowing the two of them, there's probably a picture of it somewhere.) English, as it turns out, is particularly prolific at verbing nouns or nouning verbs. Conversions from one grammatical category to another generate new meanings. In the recent history of the language, the noun *nut* (which had previously been extended metaphorically to mean "testicle") has also been verbed, perhaps via the expression *to bust a nut*, to become the verb *to nut*, meaning "to ejaculate."

And like all things, profane words eventually meet their end, ultimately fading away into banality and then obscurity. In the fifteenth century, the word *swive* (meaning "screw") was used similarly to our modern-day word *fuck*. But you've probably never heard it because it's disappeared since and in fact hasn't been used in hundreds of years, except perhaps in heated exchanges between impassioned maidens fighting over a mutton chop behind the bleachers at the Renaissance Faire.

Swive isn't the only profane word we've lost. You've probably never been called a *fart-sucker*. But that used to be a common term to describe the same thing as *brown-noser*, for reasons that you can surely surmise. And the list of now obsolete profanity goes on . . . *zounds* and *gadzooks*, which we've already discussed, as well as others, like *consarn*, which means something like *damn*, as in *consarn it!* You can see *zounds* wane over time in the chart on the next page (*swive* predates it by several centuries, where the written record is far sparser). *Zounds* had its moment in the 1800s, but by 1900, it had begun to peter out. Today, *cunt* is clearly in vogue. But this too will most likely pass, though probably not during this century.

Profanity has been in flux for all of recorded English history. And it still is. The reason relates to the effects profanity has on people and the effort it takes to maintain those effects.

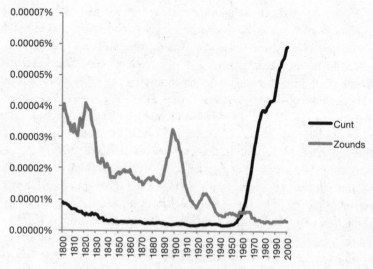

The decline of *zounds* and the rise of *cunt*.

Let me flesh that out. Using a profane word has an impact on people who have strong emotional associations with that word. But that impact weakens as a function of use. The first time you heard *shit* on television (possibly on the cartoon *South Park*), it was probably jarring—like the first time you bit into a chili pepper. But with uncensored cable television, podcasts, and social media now using the word de rigueur, your tolerance for this linguistic spiciness has increased. You might not even notice when you hear it. The more a profane word is used, the less impact it will come to have. But those words whose use is still restricted, like *cunt*, *nigger*, and to some extent *fuck*, continue to pack a punch. At least they do now. As they come to be used more, my best educated guess is that they'll fade into innocuousness and then into obscurity, just like *swive* and *zounds*.

And so, as sad as it seems, by every indication the days of *dick* and *cock* are numbered. There may well be a time in the not-too-distant future when little *Dicks* once again play in the schoolyard next to *Peters* and *Willies*, innocent to what their names once meant. At the same time, while we can't predict which they will be, we can be assured that other words will rise to fill the gaps *dick* and *cock* have left unoccupied in the ecology of the language.

8

Little Samoan Potty Mouths

There's a good chance you know what your first word was. Mine, for what it's worth, was apparently *tick-tock*. I say "apparently" because, like you, I have no personal recollection of anything that happened when I was twelve months old. You and I, like everyone else, only know what our first words were because parents and relatives have told and retold the stories of these words like the revealing pieces of our identities they're believed to be. After all, a child's first word is a critical developmental milestone. But whereas there's little to distinguish your first smile or first step from mine, a first word reveals something unique about the individual. We tend to believe that what a child says first tells us something about her burgeoning mental life, about her values or interests. A child who says *tick-tock* might be particularly interested in sounds or mechanical devices or might be in a hurry. A child who says the dog's name first (like my son did) might have a future as a biologist or might think the dog is more interesting than his parents.

So we care a lot about children's first words, especially when we are personally invested caregivers. Words for dogs and clocks are actually relatively rare first words for English-speaking children. Far more often, children first name one of their parents. A 2010 poll conducted in Great Britain found that a full 63 percent of parents reported that their children said some variant of *dada* or *mama* as their first word. And probably not in the proportions you think. In that poll, 25 percent of children were reported to have articulated a variant of *mama* (including *mom* and *mommy*) as their first word, but fully 38 percent started with some variant of *dada* (including *dad* and *daddy*).[1]

The fact that most English-speaking children ostensibly first produce the name of a parent, caregiver, or other family member makes intuitive sense to a lot of people because it comports nicely with our belief that these people are the most important parts of the child's developing universe. But interest alone can't fully account for what words a child produces first. After all, we have no reason to believe that children who say *dada* before *mama* are more interested in their fathers or love them more than their mothers. Frequency of exposure might also play a role. In those cases where a mother talks to the child more than a father does, she might say names for other people (like *dada*, for instance) more frequently than for herself. Higher frequency of exposure to *dada* than *mama* could make learning *dada* first more likely. And for that matter, maybe some words are easier than others for a one-year-old to pronounce.

All in all, it's quite hard to say why a child articulates a particular word as his or her first. And when we move beyond English, we find that it gets even more complicated. What if I told you that there's a place, an island, where children don't say *mama* or *dada* first. Nor do they say the name of the dog or an older brother or sister. And what's more, these children don't vary wildly like English-speaking children do, with some saying one word first and others another. On this island, all children say the same word first. And what if I told you that word was *shit*? That would probably change what you think a child's first word means and where it comes from.

In the late 1970s, University of California, Los Angeles, anthropologist Elinor Ochs recorded arguably the most surprising discovery ever made about how children acquire their first words, and she did it in the middle of the Pacific Ocean. Ochs was conducting research in Samoa, looking at how people there interact and use language.[2] She spent time with locals, observing their daily routines and asking about their experiences. One question she asked mothers was what their child's first word was. She doubtless expected something along the lines of patterns we're familiar with from English and many other languages: names of (human or animal) members of the household or other nouns for common objects, like *ball* or *bottle* (or, as the British survey found, *beer*). She probably also expected a lot of variability. While over half of English-speaking kids do produce a name for a caregiver first, the distribution has a long tail. When I ask students in my classes, there are usually nearly as many first words as there are students.

But when Ochs asked the mothers in the families she was working with about their children's first words, she got a completely unexpected response. Every single one of them reported the very same word. It did happen to be a noun, but it was a special one, used in a very specific way. It was the word *tae*, which, as suggested earlier, doesn't mean "mommy" or "daddy." It means "shit." More precisely, it's an abbreviation of the Samoan expression *'ai tae*, which means, "Eat shit."

This startling fact turns what we thought we knew about children's first words on its head. It means that a child's first word is determined by more than just his or her internal values. Obviously, the children in these families didn't value telling people *Eat shit* more than they valued their caregivers. No, something else must have been going on—maybe the children were exposed to specific language early on, or maybe something about the word's pronunciation made it particularly easy for them to articulate. As we'll see, nothing reveals better where first words come from and what they mean than the story of these particular Samoan children and their little potty mouths.

$ % !

Why was *shit* the first word these children said? Let's look first at the environment. Perhaps this word happened to be particularly frequent in the ambient language that the child was exposed to. Children, of course, are linguistic sponges during their second year of life. Words that you had no idea they were even paying attention to boomerang back at you in a tiny voice. And so, the words that children say often reveal, in an unsettling way, things that their parents don't notice saying. It's reasonable to presume that these Samoan children were telling people to *'ai tae* because their parents were casually tossing the expression about as well. They just got caught in the act by their little parrots.

Blaming the parents for the linguistic sins of the child is reasonable in principle, but in this particular case, the children probably didn't learn the word from their parents. When Ochs asked the Samoan mothers, they rejected outright the idea that they could be the sources. They reported embarrassment that their kids were using words that the mothers themselves shied away from. And even if we retain some lingering suspicion—perhaps the mothers simply weren't aware of what they were saying or were embarrassed to admit it to some strange woman from California or perhaps the fathers were the culprits—the story still seems unlikely. Imagine how many factors would have to line up. A derogatory expletive like *'ai tae* would have to populate the parent's speech with such overwhelming frequency that it would take pole position in the child's vocabulary ahead of other words denoting familiar people, objects, actions, or events. And this would have to be the case for every parent in every family. That's the type of parental behavior a careful anthropologist like Ochs would have noticed.

So if we think it's unlikely that parents were the sole source, then perhaps the children were learning this word from some other source in the environment. This is of course a common complaint among parents—you don't have to travel to Samoa to find parents asking where a child learned to say something vulgar. Often the answer is other kids. Could the Samoan children have been parroting not parents but older relatives and neighbors?

Again, on the surface, it looks likely. These young Samoan children interacted with other children a lot. When Ochs looked closely at how young children in this village were reared, she saw a clear difference from what she was familiar with in North America. Most American children spend the majority of their time in the care of adults—parents, older relatives, nannies, babysitters, or teachers. But in the families she observed, parents primarily cared for only very young infants. After a certain period, they recruited their older children to take over the infants' primary care. So a one-year-old might be supervised distantly by a mother and more closely by a six-year-old. Children were taking care of children.

So perhaps rather than parents, other juvenile caregivers were responsible for seeding little ears with profanity. It's not hard to imagine six-year-olds telling each other and their infant charges to *'ai tae*—especially if that was the child caregiver's own first word! The Samoan infants might have been hearing the scatological input that led to their first words not from other adults but from the children who were raising them—children who themselves may have learned to swear from the older kids who raised them, and so on.

This idea has legs. Although we like to imagine that adult caregivers provide the bulk of the linguistic guidance to young children, the fact is that children quickly reach a stage where they're learning more from other children than from adults.

Developmental psychologists Paul and Lois Bloom encountered one very clear example of this when they conducted a little experiment on their own children. They wanted to see if they could trick their kids into not swearing by surreptitiously training them to use a made-up curse word. Thus began the short and underachieving life of the pseudo-swearword *flep.* When one of the elder Blooms stubbed a toe, *flep!* Broken dish? *Flep!* Another red light? *Flep!* As Paul Bloom reports, however, it was a total failure. Whenever the Blooms cried out *flep!* their kids looked at them like they were out of their flepping minds.[3]

Once they reach a certain age, kids actually learn most of their language from peers and older children, and they do a very good job of ignoring what they hear from their parents, as psychologist Steven Pinker points out.[4] As a consequence, kids often come home with words that their parents don't use and often don't even know. They also come home using words the parents do know in ways that the parents would never imagine. Profanity is especially likely to be learned from peers, not only because it's more likely to be said on the playground than at the dinner table but also because of what kids use it for. Profanity is different from *mama* and *bottle* and other words that kids learn from their parents in that children use it as a way to show who they are—to forge their own identity. And for most kids, a lot of their identity is wrapped up in their relations with their peers. As a result, while kids learn their earliest words from their caregivers, they tend to learn later words, including taboo ones, from other kids.

But like parental swearing, kids raising kids doesn't provide a satisfactory explanation for the case of Samoan *'ai tae.* For one thing, getting the kind of consistency that Ochs observed would require an organized effort on the part of the child caregivers. Consider that merely 63 percent of British children reportedly say a variant of *mom* or *dad* first, despite the impassioned full court press that their parents apply. Certainly juvenile caregivers couldn't pull something so extensive off with *'ai tae,* with or without malice aforethought.

In sum, environmental factors—the way people speak around kids—seem like a red herring. There must be another explanation for Samoan *tae.*

$ % !

If ambient words alone didn't lead these children, across the board, to say *tae* as their first word, then perhaps the cause has to do less with what the children hear than with what they're able to articulate. Although children eventually develop mastery of the sounds of their language—by two or three years old, they're often quite proficient at repeating most of the speech sounds they hear—they don't start that way. Learning to speak is hard. Maybe *tae* is just easy for a one-year-old to pronounce.

Again, there's a lot to recommend this explanation for the *tae* mystery. So let's break it down into its component parts. First, it assumes that learning to articulate language is hard, which we know to be true. Suppose you're a year old, and you're about ready to pronounce your first word. People around you are flapping their lips, and sounds are coming out—in English, say, or Samoan—and you think this seems like an activity you could get into. But you have a lot to figure out. For one thing, you don't know what the important sounds are—which differences are worth paying attention to. The first sound of *tuck* seems different from the first sound of *truck* (the latter sounds more like what we'd spell as *ch*). Is that important enough that you need to pronounce them differently? Furthermore, you can't see much of what's going on when people pronounce words because lips and skin hide from view most of the vocal tract (the lungs, the larynx, the velum, and so on). So you have to figure out what to move when via inference and trial-and-error exploration of your own vocal tract. And finally, learning to articulate words is hard because even if you knew which sounds were important and how to produce them, you'd still have to put in the work of actually training yourself to pronounce them. You have to get the different actions right, like opening your lips and engaging your larynx, and you have to perform them at just the right time. Delay your larynx a couple of milliseconds and you might accidentally produce a *p* instead of a *b*. And all of this is just for one sound; it's even trickier to string together exactly the required sounds for a given word in the right order in real time. So we know that, as a consequence, kids systematically struggle to pronounce words early on.

The kids-say-*tae*-because-it's-easy-to-articulate story also assumes that not all sounds are equally hard to pronounce. This is also true. Some sounds come more naturally to young children than others. By about six months, infants enter a developmental stage during which they babble—producing nonsense sounds and sequences of sounds, like *dadada* or *bidubidubidu*. And the particular sounds they articulate are remarkably consistent,[5] not only within but across cultures and languages.[6] Leading theorists have even proposed that babbling, including the specific sounds children articulate, is part of a universal, genetic program children are born with and that babbling, like secondary sexual characteristics or menopause, is an automatic part of the individual's natural, scheduled maturation.[7]

The early sounds that infants around the world produce during babbling won't surprise anyone who's spent time around children. As far as consonants go, the ones you typically find in babbling include *m*, *n*, *b*, and *d*, and these are often followed by *p*, *h*, *f*, *t*, *k*, *g*, *f*, and *w*.[8] As for vowels, across languages, children seem to be more proficient early on with vowels that you might transcribe as *ee*, *ey*, *uh*, and *ah*.[9]

Of course, as there is for just about every claim regarding innate predispositions driving human development, there's conflicting evidence on this. For example, deaf children were once believed to babble just like hearing infants—which would be a pretty bulletproof piece of evidence for a universal genetic basis for babbling.[10] But more thorough recent studies show that deaf children don't start verbal babbling until about ten months, several months after their hearing counterparts.[11] At the same time, deaf infants do display early manual babbling—repeated and stereotyped language-like movements of the hands.[12] The fact that deaf children babble manually, together with the delay in their verbal babbling, suggests that input matters for the babbling that children do. What's more, other evidence demonstrates that the language surrounding an infant can shape, if subtly, the sounds he or she produces.[13]

Nevertheless, early speech sounds—at least in hearing infants—are relatively predictable. And so are the syllables that children assemble those sounds into. As I discussed in Chapter 2, syllables are the rhythmic units that structure words. And they aren't all equally easy for children. Syllables come in different types. In the simplest terms, you can think of any syllable as structured around one vowel and optionally one or more consonants that precede and/or follow it. So if we write out consonant and vowel sounds as C and V respectively, then *the* is a CV syllable. (Importantly, *the* isn't CCV because we're focusing not on spelling but on pronunciation. The word *the* has just one consonant and one vowel because the two letters *th* indicate a single consonant sound, produced by putting the tongue between the teeth to create a noisy baffle for air passing through. In Old and Middle English, in fact, this sound was written with just one single letter, þ.) The simplest syllable is a V syllable, like the words *a* or *I*. Children are pretty good at V syllables. They're even decent at VC syllables, like *it* and *up*. But early on, they seem to prefer CV syllables, which are in evidence in words like *mama* and *dada*, each of which just repeats a single CV syllable twice.[14]

We know that CV syllables are the young child's go-to because even when trying to imitate more complicated words, they often reduce them to fit this frame. A commonly encountered example is *spaghetti*, which rendered in syllables is CCVCVCV, as you can see on the next page.

s	p	a	gh	e	tt	i
|	|	|	|	|	|	|
C	C	V	C	V	C	V

That first syllable *spa* often trips children up, and they get around it by simplifying the word in their pronunciation. Often they omit a sound (*pa-ghetti*) or even a whole syllable (*ghetti*). Reductions like this are so import-ant for us because what the child says differs from what he or she hears. That tells us what the child can and can't do at a given age. Generally, at around a year of age children are good at CV, V, and VC syllables, and they're decently proficient at a range of sounds.

So we've established that learning to pronounce speech sounds is hard, but not uniformly so. Children follow largely similar pathways in develop-ing their first sounds and syllables. This means that they're more likely to articulate certain word-like sequences of sounds early in life and to sim-plify words of the ambient language in particular ways. Could it be that *tae* just happens to be assembled from easy, typically early sounds?

To answer this, we need to break down the word. The *t* seems like a good candidate for early children's speech. But it's not pronounced how you think it is. Spelling can be deceiving. Samoan words spelled with *t* are only pronounced like the English *t* in formal speech. In informal reg-isters—for instance, when telling someone to '*ai tae*—it's pronounced like *k*.[15] (If that seems unreasonable, consider how much stranger English or-thography is by comparison. Compare how *t* is pronounced in *tree*, *nation*, and *the*.) So the word begins with *k*, a sound that children around the globe are a little slow to master. *K* doesn't usually come online until after *m*, *n*, *b*, and *d*. Then comes the *ae*. This is a diphthong. You should be familiar with diphthongs from English, which has a lot of them. They're just vowels that start in one part of the mouth and end somewhere else. For instance, the vowels of English *night* and *out* are diphthongs. The diphthong in *tae* is similar to the vowel of *night*, but it ends with the mouth a little wider open. The upshot is that it's a complicated sound. Not exactly what you'd expect in a first word.

In sum, *tae* seems like it would be moderately challenging for a child to pronounce. The *t*—that is, the *k* sound—is not well represented among children's earliest sounds, although it does come along soon after. The diphthong vowel itself would not be easy. So it's a stretch to think that chil-dren would regularly articulate precisely *tae* as part of their exploration of speech sounds. And even if they did, this would still be only one of many syllables they'd be articulating and typically not among their first.

So the question remains, why is this rare, somewhat challenging syllable interpreted as the Samoan child's first word? The answer will bring us back to the parents. After all, English-speaking infants presumably also articulate something like this same syllable at points. And yet English-speaking parents rarely record their children's first words as *caw* or *kite*. Although we originally exonerated the parents for *tae*, at least in modeling behavior, we have to return to them now to understand why, in the face of a flood of word candidates during babbling and then single syllables, they interpret this particular sound as their child's first word.

$ % !

Suppose that children across the world and across languages produce largely similar patterns of sounds through babbling and early word attempts, modulo some influence of the ambient native language. In that case, several factors will determine what a caregiver counts as a child's first word. Not least of these are, of course, the words of the caregiver's language. If your child produces something that sounds like *dada* and your language has a word that sounds something like that, then there's a chance you'll interpret it as that word.

But perhaps more important are your expectations about what sorts of words a child is likely to say. If a child says something like *dada*, a caregiver could interpret that as an approximation of *dada* or *daddy*. But because infants' word-like vocalizations at around one year don't quite use the specific sounds of the language around them, it might sound less like the father label *dada* and more like *Dada*, the European avant-garde art movement of the early twentieth century that's pronounced slightly differently. The same vocalization might sound a lot or even more like *dawdle* or *ta-ta* or *duh-duh*. A caregiver might assume that this ambiguous sequence of sounds is an incipient approximation of *dada* in part due to the belief that a one-year-old will have no particular interest in or knowledge of Dadaism or dawdling. No, the caregiver presumes that the child is interested in Dad.

And these two factors are causally related. Paradoxically, the assumptions of the caregiver affect the words that a language has to choose from. Let me explain.

Language	"Mother"	"Father"
Swahili	mama	baba
Kikuyu (East Africa)	nana	baba
Xhosa (South Africa)	-mama	-tata
Tagalog (Philippines)	nanay	tatay
Malay	emak	bapa
Romanian	mama	tata
Welsh	mam	tad
Urdu	mang	bap
Turkish	ana, anne	baba
Pipil (El Salvador)	naan	tatah
Kobon (New Guinea)	amy	bap
Basque	ama	aita
Hungarian	anya	apa
Dakota (United States)	ena	ate
Nahuatl (Mexico)	naan	ta'
Luo (Kenya)	mama	baba
Apalai (Amazon)	aya	papa
Chechen (Caucasus)	naana	daa
Cree (Canada)	-mama	-papa
Quechua (Ecuador)	mama	tayta
Mandarin Chinese	mama	baba

In Chapter 1, we saw that most words in spoken languages are arbitrary, with the exception of sound symbolism (or onomatopoeia). But I glossed over another exception at the time. The world's languages show remarkable systematicity in their labels for parents. Consider the above list (compiled by Larry Trask in a delightful essay on the topic) of words for mom and dad in a host of largely unrelated or distantly related languages.[16]

Nearly every word for mother in this list has an *m* in it, and those that don't have an *n*. Words for father are more variable but still use the same collection of sounds that we're familiar with from babbling: *b*, *p*, *d*, and *t*, along with the vowel *a*. And these trends are representative of the world's languages. In 1959, anthropologist George Murdock published a very large survey of terms for parents in nearly five hundred of the world's languages.[17] More than half (52 percent) had words for mothers that used *m* or *n*. Words for fathers again showed more variability, but more than half (55 percent) included one of four syllables: *pa*, *po*, *ta*, or *to*. This is clearly not random. Some process or processes must be conspiring to populate the languages of the world with words for mothers and fathers that sound similar.

The explanation isn't sound symbolism: mothers don't sound more or less like *m* or *n* than any other thing that a word could name. It's also not due to some sort of linguistic founder effect, whereby the words used by humans 100,000 years ago have persevered to this day. That would be impossible. Given how quickly words are replaced in the languages of the world, words spoken at a time before all languages diverged from one another (if indeed that's what happened, which the jury is still out on) would be totally unrecognizable by now.

No, the languages of the world have *mama-* and *papa-*like words because parents hear what they want to hear.

Influential linguist Roman Jakobson—an exile of the Bolshevik revolution in Russia who founded the influential Linguistics Circle of Prague in the early twentieth century before he was again exiled during World War II and found his way to a chair at Yale—explained this better than anyone has since.[18] Parents, he argued, have expectations about what their children will say, and they specifically expect that their children will talk about them. Since mothers throughout history—for biological and cultural reasons—have tended to spend more time with infants than fathers, they've often been the closest on hand to deploy that assumption, and as children most commonly produce something like *mama* before *dada* or *papa* (contra the findings of the recent British survey), *mama* gets interpreted as referring to the mother.

This is especially true when a language already has a word like *mama* for mother, but it's also true even when that's not the case. Parents are quick to allow for deviations between how adults use language and how children approximate it. So it's common for caregivers to accept children's deviant words and pronunciations as proxies for adult equivalents. If I expect my child to call me *daddy*, but he's saying something like *dahdah*, I'll take it. But parents go farther. They don't just categorize what the child says as an instance of an existing word; they frequently repeat, reuse, and reinforce it. At fourteen months, my son couldn't (or just didn't) pronounce the *t* sounds of *night-night*. (Notice that he predictably changed CVC syllables to CV.) So the rest of the household copied him and told him it was time to go *nigh-nigh*. At fifteen months, he couldn't say *lifeguard*, and we followed his lead in referring to the *gar-gar* at the pool. Adults come in this way to adopt changes originated by children that the caregivers interpret as meaning specific things. As a consequence, the language changes. When enough adults use a word, the word effectively becomes a word of the language. And by this process, even if a language doesn't have a word like *mama* to start with, it can gain that word in a generation.

This explains why so many languages have such similar words for mothers and fathers. And it also explains why, even when they're dissimilar, words for mothers and fathers are simple for children to pronounce.

They're simple because of the interplay of evolution—which has endowed children with brains and bodies that are similar in their abilities and their proclivity to pronounce certain sounds at certain stages—with cultural expectations that caregivers apply to their children and with cultural change on the time-scale of years and decades, during which changes to a language take hold in a community.

And so, with that in mind, we can return to Samoa. Samoan does have words for mother and father, *tina* and *tama*, respectively (pronounced *kina* and *kama*). But Ochs's study reported neither of these as a child's first word. That, of course, was *tae*. So why doesn't Jakobson's explanation apply here as well? Why would mothers in Samoa not believe what their counterparts around the world do—that their children's first interpretable utterances refer to them? It came as a surprise to me, though perhaps it wouldn't to anthropologists more familiar with how cultures can vary around the world, that the Samoan parents Ochs was working with had specific beliefs about children that led them down a different logical path.

Here's the way the Samoan villagers whom Ochs worked with thought about children, as explained by Ochs herself:[19]

> From the Samoan point of view, the small child is heavily under the influence of *amio* [natural drives that lead people to act in socially destructive ways]. Infants and small children carry out such outrageous behaviors as running and shouting during a church service or formal chiefly council meeting, throwing stones at caregivers, hitting siblings and the like, because they are [believed to be] incapable of . . . suppressing *amio*.[20]

These Samoan parents considered children uncontrollable, unruly, and socially destructive. It's not hard to see what would lead someone to such a belief. I have thirty pounds of insuppressible *amio* at home myself. The difference between these Samoan mothers and, say, my household isn't that Samoans think children are out of control, whereas we don't. It's in the balance between this set of beliefs and the idea that children are incipient little communicators, seeking engaged interactions with another person. My family has a little more of the latter, in the balance, and the Samoan mothers, more of the former. And if you're more prone to think children are ruled by *amio*, then your expectations about what they're likely to say will reflect this belief, just as your belief that children are trying to connect through language with their parents will create your expectation that they'll say a caregiver's name first. Uncontrollable, unruly, socially destructive little people are more likely to tell those around them to 'ai tae. And so, whereas an English-speaking North American might interpret something a child emits that sounds like ka as car or cat, the Samoan parents in question were apparently led by their expectations to believe that the child's little amio was hurling an antisocial expletive.*

And so ends the mystery of the diminutive Samoan swearers. Their parents were exceptional not because they happened to have diabolical children or because they let themselves incautiously swear in the presence of infants. Like caregivers around the world, they let their beliefs affect their expectations. In the mushy, imprecise vocalizations of their children, they heard what they anticipated hearing. In other words, in this particular case, the kids didn't have potty mouths; the adults had potty ears.

---

* The Samoan mothers told Ochs something else, which she was kind enough to relay to me. They told her that, at some level, they liked their kids to be tough, for self-protection. Sticking up for yourself can be a useful survival tool. And so, as they explained it, "sometimes bad is good." This belief might have contributed to making the parents more prone to hearing profanity.

9

Fragile Little Minds

In the summer of 2014, Danielle Wolf was arrested after swearing in front of her two children in a North Augusta, South Carolina, supermarket. Accounts of the event differ. Everyone agrees that she said, "Stop squishing the fucking bread!"[1] But there's some debate over whom she said it to. Wolf claims she uttered it to her husband, who she said was throwing frozen pizzas into their shopping cart, crushing the bread beneath. A witness claims Wolf said it to her children. Regardless, by all accounts the children were within earshot. The witness called the police, and Wolf was arrested for disorderly conduct. According to a North Augusta city ordinance (§ 16–88[12]), disorderly conduct includes "utter[ing], while in a state of anger, in the presence of another, any bawdy, lewd or obscene words or epithets."[2]

And lest you think this is just a peculiarity of South Carolina (not that I'm suggesting there's any reason to expect aberrant behavior in South Carolina),[3] something very similar happened in my own backyard, in Southern California, in 2006. Elizabeth Venable, a PhD student at the University of California, Riverside, was at the John Wayne Airport when she was overheard using profanity with a friend. As the *Los Angeles Times* reports it, a sheriff's deputy, "noticing several families with small children nearby," asked Venable to "please watch her language while at the airport."[4] According to the *Orange County Register*, Venable asked, "Is it against the [expletive] law to say [expletive]?"[5] Although the paper didn't report what the expletives were, we can make an educated guess. That's when the deputy cited her. Later she was charged with two misdemeanors: disorderly

179

conduct and disturbing the peace. The rule she violated appears to be an Orange County ordinance that applies specifically to John Wayne Airport and prohibits people from being "disorderly, obnoxious, indecent or commit[ting] any act of nuisance." Given how strongly he objected to profanity in film, I suspect the Duke himself would have approved of this enforcement of the ordinance in his eponymous airport.

To be clear, in neither of these cases was there a question of harm being done to anyone, including children, beyond direct or indirect exposure to profanity. Wolf was not charged with physically or verbally abusing her children; nor was Venable accused of threatening children in the airport terminal. The deed that landed them in court was using the word *fuck* around children. There's a lot to say about whether laws like these are constitutional, especially with respect to the constitutionally protected right to freedom of speech. We'll pick up that thread later in the book. For now, I just mention these cases to highlight our deep-seated belief—entrenched in our legal codes and our social practices—that exposure to profanity harms children.

If profanity causes harm, we should care. Society, its public institutions, and we as individuals all have a responsibility to ensure the well-being of children. Do we need to monitor the language our children are exposed to in order to protect them? If profanity really does hurt children, there should be social and legal consequences—swearing around children might be most appropriately considered a form of child abuse and should be punished accordingly.

The American Academy of Pediatrics (AAP) appears to have the answer. In the fall of 2011, it issued a press release warning that exposure to profanity is harmful to children.[6] The AAP was relying on a recent study from the prominent and respected medical journal *Pediatrics*. The AAP's statement had two parts. First it announced that being exposed to profanity increases children's use of profanity—in the case of the study, middle school students. That by itself probably wouldn't surprise many people. We know that kids pick up on what they hear. But the AAP then went on to report that swearing by minors was "a risk factor for increased physical and relational aggression." Young minds, the AAP would have us believe, are fragile—so fragile that even profanity can damage them.

The media picked the story up quickly and, as they will, amplified it in the retelling. For example, the *Daily Mail* quoted a media expert who interpreted the results as indicating that "children who use profanity are more likely to make them more aggressive towards others [*sic*]."[7] *Medical News*

Today reported the findings like this: "Bad language and profanity . . . appear to increase aggression in teenagers."[8] And one article cited the first author of the study in question explaining how it works: "Profanity is kind of like a stepping stone. . . . You don't go to a movie, hear a bad word, and then go shoot somebody. But when youth both hear and then try profanity out for themselves it can start a downward slide toward more aggressive behavior."[9] The party line is clear: profanity is dangerous because it's a gateway form of aggression.

And this is just one way in which profanity has been argued to harm children. The *Pediatrics* article identified several other possible dangers that profanity poses.[10] First, "exposure to profanity can induce a numbing effect on normal emotional responses." And second, exposure to profanity causes people to "experience negative physiologic responses, such as increased heart rate or shallow breathing."

If true, these claims are a big deal. For instance, if they're genuine, then you probably shouldn't let your twelve-year-old read this book.

So let's take a close look at the *Pediatrics* article's argument. To reiterate, it raises three core issues:

1. Exposure to profanity can cause direct harm to children.
2. Exposure to profanity makes children more likely to use it.
3. Using profanity in turn makes children more likely to act aggressively.

Because this article appeared in the prestigious, peer-reviewed journal *Pediatrics*, it's tempting to take these claims at face value. But not everything you read withstands the weight of careful scrutiny. So let's take a look, as informed consumers of science, at each of these points. How compelling is the evidence? What do we need to protect children from and why?

$ % !

Point one: Does exposure to profanity directly harm children?

The original 2011 *Pediatrics* article describes two direct effects of profanity—a numbing effect on normal emotional reactions and an increase in immediate negative physiological reactions. Let's deal with both of these in turn.

First, numbing. The paper claims, "Studies have found that exposure to profanity can induce a numbing effect on normal emotional responses."[11]

This is a pretty bold claim. Where does it come from? The paper itself doesn't actually demonstrate this—it's part of the article's review of the existing scientific literature. But if you follow the references, the *Pediatrics* article only cites a single study that ostensibly demonstrated emotional numbing, a 1989 article titled "Desensitization to Television Violence: A New Model."[12] Here's the problem. That article doesn't mention profanity— at all. It doesn't contain the word *profanity*. Or even the word *word*. Or *language*. To its credit, the paper is quite interesting: it offers a new theoretical explanation for the finding that exposure to prolonged "scenes of violence," like "primitive mutilations," over time leads to decreased emotional reactions in viewers. That may well be true! But the article doesn't talk about language at all. And it only talks about adult viewers—no mention of children. Here's the final nail in the coffin: the desensitization article is completely theoretical, so it presents no new data and has nothing to do with language. Nor do any of the twenty-three other articles that it cites. I checked.

You might reasonably be wondering why the *Pediatrics* study didn't cite more relevant research showing that profanity numbs emotions. For instance, what about studies demonstrating empirically that profanity desensitizes people to violence? Those articles were not cited because they simply do not exist. There's no evidence of the numbing effects of profanity—not for adults, not for children. Now, I'm not the first person to note that absence of evidence is not evidence of absence. It could well be that exposure to profanity does in fact directly desensitize people to violence and that scientists simply haven't asked the question in the right way. But at present, there's no empirical evidence that exposure to profanity dulls children's reactions to violence. None. The *Pediatrics* article got that wrong. Good thing we checked.*

The *Pediatrics* article further claimed that profanity affects well-being by leading people to "experience negative physiologic responses, such as increased heart rate or shallow breathing."[13] This is sort of the opposite of the initial claim, on a shorter scale. It asserts that profanity causes harm in the moment of exposure because people's bodies react strongly to it. The authors cite no evidence for this claim. But a glance at the literature on the effects of exposure to profanity reveals that the *Pediatrics* article is half

---

* I don't mean to sow mistrust in science generally. But scientists are humans, and humans sometimes get things wrong. So it's not unreasonable to take a page out of Ronald Reagan's book: trust, but verify.

right. There is indeed a lot of evidence that the body reacts to profanity in predictable ways, including increased heart rate and shallow breathing. As you might recall, we discussed this in Chapter 5. For example, one 2006 study measured people's heart rates while they read neutral words, school-related words, unpleasant words, and taboo words.[14] Heart rates started going up about two seconds after participants saw a taboo word and kept going up for several more seconds.

But does an increase in heart rate mean, as the *Pediatrics* article claims, that people were experiencing a "negative physiologic response"? Not at all. Increased heart rate and shallow breathing both accompany an assortment of fundamentally quite different states. Heart and breathing rates increase when people perform a hard mental task, like solving an arithmetic problem or playing a computer game; they also increase when people are emotionally aroused, whether angry or fearful or overjoyed.[15] They increase in response to not just taboo language but pretty much anything that's arousing. (The word *arousal* has a specifically sexual connotation in English, but psychologists use it for responsiveness to stimuli in general.) For instance, a study of college students showed that heart and respiration rates both increased during texting, especially while students were sending and receiving a message[16] (perhaps because they were in a more aroused state when anticipating how their message would be received or what they would find in the response). The point is that these two physiological changes provoked by profanity aren't intrinsically negative. The problem with the *Pediatrics* paper's claim lies not in the facts about the immediate reaction profanity causes in your heart and lungs but in the interpretation of what that reaction means. It means only that people become emotionally aroused when you bring them into a lab and present them with profanity. That's not surprising. But this in no way implies that profanity causes harm—any more than sending text messages or doing mental arithmetic does.

The closest thing to profanity causing harm that you can find in the literature—and you really need to dig for it—comes from public health studies. But, as we'll see, even those don't directly address the question of profanity. A number of these studies ask how different types of caregiver abuse predict children's psychological and physical health outcomes. For instance, take a large 2010 study that surveyed nearly ten thousand Scandinavian teenagers. It asked them whether they had experienced certain types of abuse in the previous twelve months. Aside from physical abuse, they were also asked about verbal abuse. And this category included

questions about "sulking or refusing to talk, insulting or taunting or swearing, throwing objects and threatening with violence."[7] Teens who reported
exposure to at least one of these behaviors were pooled together as subject
to "verbal abuse," and as a group they were more likely than those who
reported no abuse of any kind to also report psychological problems like
depression, anxiety, aggression, and hyperactivity.

Surveys like this one focus our attention—appropriately—on how verbal abuse might affect the psychological well-being of adolescents. There's
ample evidence that abuse correlates with psychological problems. This
study showed that verbal abuse, defined in the study's specific way, correlated with an increase of about 16 percent in aggregate psychological
difficulty score—a measure of the psychological problems the adolescents
were having. Far more damaging was physical abuse, which correlated
with an increase in this score of 37 or 49 percent, depending on whether
the abuse was classified as "mild" (e.g., pushing, shoving, shaking, or whip-
ping) or "severe" (e.g., battering, hitting with objects, or kicking). I point
out the substantial effects of physical abuse not to marginalize the effects
of verbal abuse but just to highlight the different things known to correlate
with negative psychological health outcomes in minors and their relative
severity. Different types of abuse hurt children to different degrees.

According to this Scandinavian survey, profanity is not demonstrably
one of those types of abuse. Studies like this one don't tell us anything about
profanity per se because all such studies slip profanity in with things that
are clearly verbal abuse and don't allow us to tease them apart. In this case,
profanity was lumped together with things like threatening with violence,
throwing objects, and insulting the child, for instance. And any participant
who said yes to any of the verbal abuse questions was categorized as having
been subject to verbal abuse. So there's no way to tell whether these other
activities—which clearly constitute abuse—were responsible for the entirety
of the effects observed or whether profanity also played a role.

And there's good reason to suspect that if you were to split out exposure
to profanity, you wouldn't see any effect on child well-being at all. Profanity
can, of course, be used for many different purposes. It's absolutely true that
some of those are abusive. But it stands to reason that calling a child worth-
less or threatening physical harm is likely to increase the child's depression
and hostility, whether profanity makes an appearance or not. Conversely,
it's hard to imagine what harm is likely to come from telling an adolescent
his report card is *fucking fantastic*. Compare that with telling him *You're*

stupid, and I hate you. You can use profanity positively, and you can use "clean" language abusively. Profanity is orthogonal to abuse.

Verbal abuse of children is, of course, an issue we should be studying more and raising vigilance about. But there's no evidence either that profanity itself does direct harm to children or that abusive language is more harmful when it includes profanities. Again, I have to raise the specter of the absence of evidence. We know only that we don't have evidence that profanity hurts children directly. But in the absence of that evidence, one would also expect an absence of the pretense that evidence exists.

$ % !

So we've debunked the *Pediatrics* article's first claim: that exposure to profanity harms children directly. Let's move on to the second claim: that the more children are exposed to profanity, the more they will come to use it themselves.

In the limit, a categorical version of this claim must be true. Children learn language based on what they hear around them. That usually includes profanity. So if they don't ever hear profanity, they won't learn it. And although many parents try to shield their children from profanity, this is largely impossible. Parents often try to self-censor when a child arrives, but this can be challenging, as anyone who has stepped on a Lego block in bare feet or been micturated upon at the changing table can attest. Not every swear gets stifled.

And, of course, children learn language not just from their primary caregivers but also from other speakers of the language around them. And these other people often don't buy into the idea—an aunt, grandfather, older sibling, or friend might not see any problem with an occasional *shit* here or *bastard* there. Maybe one of those individuals even secretly—or not-so-secretly—wants to sabotage the whole "linguistic whitewashing for the children's sake" enterprise. So children are going to be exposed to profanity by people around them.

And also: the Internet.

But the article proposes a more nuanced, quantitative relationship. Does more frequent use of profane words around a child lead that child to use them more? Maybe.

There's a long history of studying how a word's frequency in a child's environment affects acquisition of that word. Children typically begin to understand a few words in their first year of life and utter their first words around their first birthday, as we saw in the last chapter. From there, they'll typically acquire a few words per month until sometime around eighteen months (give or take six months), when their vocabularies begin to explode in a "word spurt." By the time they start school, their vocabulary will have ballooned to something like 10,000 words, and by college, it will be five times larger.

And it's reasonable to think that the words children hear most frequently are also those that they learn earliest and use most. But it's complicated. Children's early words tend to name people, animals, things, parts of things (especially body parts), and actions, or to mark social interaction routines (like *bye-bye*, *up*, or *all gone*). The distribution of these sets of words in a child's early lexicon varies between children and across languages and cultures—English-speaking kids learn more early nouns than their Korean-speaking counterparts, for instance.[18] And while many of a child's first words are frequently occurring ones (the name of a family pet or a word for a commonly experienced object, like a doll, ball, or bottle), a child's earliest words often do not include the most frequent words in the language. For example, the ten most frequent English words (as measured by the spoken component of the American National Corpus)[19] are, starting with the most frequent, *the*, *of*, *and*, *to*, *a*, *in*, *it*, *is*, *for*, and *I*. Yet none of these are commonly found in a child's earliest words. In fact, if you look at caregiver speech directed at children ("child-directed speech"), you find that more frequent words are learned slightly later than less frequent ones, in large part because more frequent types of words (like articles, prepositions, and so on) are learned later than less frequent words like nouns.[20]

So at the very least, we can say that the story is more complicated than merely frequency influencing word learning. In fact, when you split words into different groups—common nouns, words describing people, action words, and so on—you do indeed find that within each category the more frequent words tend to be learned earlier. For example, here's a graph of the age at which one particular child first used each of his nouns (his age is on the *x*-axis) plotted against how frequent that word was in the child-directed speech he heard (it's actually the log of word frequency because frequency effects in language have logarithmic effects).[21] You can see that within nouns, the child learns more frequent ones earlier, on average, and then moves on to learn less frequent ones as well.

Nouns

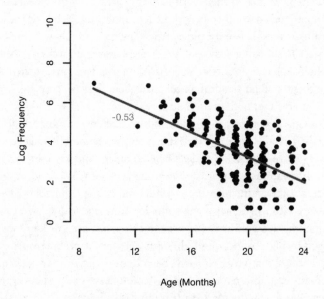

Each dot represents the first time the child produced a particular noun; more frequent nouns tended to be learned earlier than less frequent ones. Image reproduced from B. C. Roy et al. (2009), used with permission.

Of course, a reasonable person could object to studies like this one. Correlation does not imply causation. So the fact that children tend to learn more frequent words earlier doesn't entail that frequency is the reason for earlier word learning. Other factors might be in play. For instance, more frequent words are shorter, all things being equal. And children learn shorter words earlier. Maybe frequency plays no causal role.

To know for sure, you'd need to run an experiment: you'd have to manipulate how often children heard particular words and see whether this factor alone, holding all other possible causes constant, affected children's learning of the words. There have been several such studies. Perhaps the best known took place in the early 1980s.[22] Researchers interacted with one- to two-year-old children over the course of four months and, among other things, introduced them to some made-up words, like *tete* or *fus*.

The words had to be made-up because otherwise the children might encounter them by chance in their normal lives as well. And the researchers carefully manipulated whether, over those four months, the children heard each word ten times or twenty times. And they measured how likely the children were to spontaneously use those words. The result was clear: they learned the frequent words relatively often (on average, kids learned 44 percent of them) and far better than the infrequent ones (29 percent). This effect was observed for both nouns and verbs.

So it seems that a causal relationship exists between how frequent words are in a child's environment and when he or she will start using them. But how well does this scale? Once an older child knows a larger, more adultlike number of words, say on the scale of tens of thousands, does the frequency with which those words entered the child's ears affect the frequency with which they come out of his or her mouth?

We don't have a conclusive answer to this question. We don't even know exactly what proportion of children know what profane words at what age, because that research hasn't been done. Typical surveys asking parents about their children's vocabulary tend to leave out profanity, and parents usually aren't particularly forthcoming in offering swearwords that their children know. The one source we have is an observational study, in which a set of researchers spent a year recording all taboo words uttered by children around them—predominantly middle-class kids from the northeastern United States.[23] The results showed that the swearwords children use change with age. The youngest English-speaking children are more likely to use words like *poopy* or *stupid* with the force of profanity, whereas by adolescence they're using more adultlike profanity: *fuck*, *shit*, and so on. We also know that boys and girls swear differently. *Oh my God* ranks as the most frequent profane expression among girls aged one through twelve, whereas boys in the same range are most likely to use *shit*.

So we have some hint as to when children use which profane words, but we certainly have no evidence of how frequently they use them. The only exception is the *Pediatrics* study. And that evidence is rather weak, as we'll see.

$ % !

Let's look at exactly what the *Pediatrics* study did. The researchers asked students from a large midwestern middle school a series of questions. To get a measure of how much the children were exposed to profanity, they asked the kids to list their three favorite television programs and video games, then rated these for how much profanity they contained. To determine what the children thought about profanity, the researchers asked them to give their opinion about statements like "I think it is okay for me to use profanity in my conversations" on a scale from 1 (never true) to 5 (almost always true). They also wanted to know how often the children actually used profanity, and to measure this they asked them to rate a series of statements like "I use profanity in my conversations with my friends" from 1 (never true) to 5 (almost always true). And finally they wanted to know how aggressive the children were. So they asked them to rate a number of statements describing physical aggression ("I hit, kick, or punch others") or relational aggression ("I have tried to damage [a] person's reputation by gossiping about that person"), again on a scale of 1 to 5.

With these data secured, the researchers dug into them to discover any significant correlations among the various measures. And some of them did correlate. In particular, children who reported being exposed to media with more profanity were also more likely to report that they found profanity acceptable. In turn, those who found profanity more acceptable were also more likely to report using profanity. And those who reported using more profanity were also more likely to report acting aggressively. The article interprets this chain of correlations as implicating profanity in aggression: observing profanity affects attitudes about words, attitudes affect the use of those words, and word use affects aggression.

But not so fast. Can a study like this actually tell you whether profanity causes this chain of effects?

Let's tease them apart, starting with the first couple of steps. Right now we're focused on the question of whether exposure to profanity increases its use. The study states that adolescents who reported watching shows and playing games with more profanity in them also reported finding profanity more acceptable and using more profanity themselves. Does this answer the question about frequency? Does this mean that exposure to more profanity leads to more use of profanity?

We don't know, because the study was correlational. It's not always obvious why correlation doesn't imply causation, so let me just remind you here. (If this is old hat to you, by all means, skip to the next paragraph.) Here's a nice example of why you can't infer causation from correlation.[24]

Suppose you want to know whether religious faith causes an increase in al-
cohol consumption. You might try to find an answer by counting the number of bars and the number of churches in each of a large number of US
cities. You'd surely find a significant correlation; overall, cities with more
bars have more churches. But what's the right causal interpretation of these
data? That religious practice causes drinking? Maybe. I certainly know
people who claim to need a drink after church. But the reverse causality is
equally possible: maybe increased drinking causes more religious practice.
You can come up with a story for this too. Maybe when people drink more,
they do more things that they feel they need to atone for. And even worse,
it's equally possible that something totally unrelated to churches or bars—
some hidden variable that you didn't even consider—causes concurrent
increases in both. For example, maybe cities with more churches also have
more bars because those cities have larger populations in general. In short,
a correlation between two things cannot tell us whether they're causally related or, if they are, in which direction. That's true whether we're interested
in the number of churches or exposure to profanity.

So let's apply this reasoning to the *Pediatrics* study. The data about exposure to profanity, attitudes toward profanity, and use of profanity were
all self-reported. So we know that children who reported exposure to media
with more profanity also reported more positive attitudes toward using
profanity, and in turn those who reported more positive attitudes toward
profanity also reported using more profanity. You might be able to spot the
hidden variables here. The problem is that all the data rests on the children's
reporting. We don't have any idea of their actual behaviors, beliefs, or view-
ing patterns. And these children's reports could differ from reality for many
reasons. Perhaps some of the children intentionally misled the researchers.
For example, some children could have been afraid to confess that they
were exposed to, had positive attitudes toward, and tended to use profanity.
Imagine yourself as a middle school student. A team of researchers from
prestigious Brigham Young University shows up asking about profanity and
aggression. You might think twice before giving truthful answers. Some
children might be more honest than others. And if the less honest children
artificially deflated their scores on each of these questions, this would produce the precise correlation that the study found. It would create the false
impression of correlations in experience and behavior where the real effect is
just a tendency for some children to respond less truthfully across the board.
Note that it could also work in the other direction: some children could have

been boasting about both the shows they liked to watch and the amount they swore. This too would create a spurious correlation in exactly the same direction: children who reported watching more profane media would boast of more positive attitudes toward profanity and more profanity use.

This is just one of many possible hidden variables that might describe the actual causes underlying the correlations that the study observed. For instance, maybe children with caregivers who don't believe there's anything wrong with swearing both allow their children to watch media with profanity and are more likely to encourage or permit its use.

And there's also the problem of reverse causation to deal with. It could very well be, as the article suggests, that exposure to profanity causes increased use of profanity. But what about the opposite direction? Maybe children who swear are more strongly attracted to profane television shows because such programming sounds more like what they sound like or what they want to sound like.

So does increased exposure to profanity lead to greater use of profanity? The noun and verb research discussed earlier suggests that it's certainly possible that children exposed to more profanity will learn it earlier. But that doesn't imply that they'll use it more, and the correlational study we've just been discussing is causal Swiss cheese. However, just to see where the argument goes, let's suspend disbelief and stipulate, for the moment, the possibility that children will use profanity more the more often they're exposed to it. The really important part of the story is whether using more profanity causes harm to children. That's next.

$ % !

The *Pediatrics* study discovered a correlation not just between profanity exposure and use but also between how much children reported using profanity and how much they reported engaging in both physical and relational aggression. This part of the study suffers from the very same problems we just identified. For one thing, there's still the hidden variable caused by self-reporting. As we saw before, correlations between self-reported variables might tell you more about how the respondents deal with the survey than about the behavior it asks them about. People in a study like this might not tell the truth for a variety of reasons. Adolescents who were more candid about aggression might also have been more candid about the media they were exposed to.

Many other possible hidden variables could be behind the correlations between profanity use and aggression. Which of these seem plausible to you? Maybe children exposed to more profanity also happen to be exposed to more violence in media, which causes them to be more aggressive. Or maybe children exposed to more violence in real life both seek out media with more profanity and also act out more aggressively. Or maybe children who come from families that are more predisposed to aggressive behavior for any of a number of reasons (medical, cultural, socioeconomic, genetic, and so on) both seek out more profane media and also exhibit more aggression. I'm not committing to any of these as the single, real cause of aggressiveness. But if you think any of these are plausible, and actually even if you don't, they reveal the hidden variable problem inherent in correlational studies. People come with baggage. And some of that baggage might be causally related in a systematic way to the things you're interested in.

The people who do correlational studies cleverly try to control for those other factors by measuring and including them in their models. So if you believe that exposure to violence in media could be a confounding factor—it correlates with exposure to profanity and could explain some amount of aggression—then you measure not only how much profanity but also how much violence children are exposed to. The two will probably correlate, but the key point is that you can measure exactly how much media violence correlates with child aggressiveness, and you can pull that apart in a statistical model from the amount that profanity exposure correlates with child aggressiveness. The authors of the *Pediatrics* study tried to do this. But to know that profanity exposure per se and not any of these other possible confounding factors is responsible for increased reports of aggressiveness, you'd need to do the same thing not just for exposure to media violence, as the authors did, but for every other possible confounding factor, which they did not. That means that the real culprits—the real causes for increased aggression, if in fact there are any—may still be out there.

And once again we have to deal with the possibility of reverse causation. Suppose we find that people who swear more really are also more aggressive. Which do you think is more likely? That using more profanity causes people to be more aggressive, as the article posits? Or that being more aggressive causes people to use more profanity? I would bet on the latter. In fact, there's even some experimental evidence of this. A study conducted at Keele in the United Kingdom asked whether feeling more aggressive would cause people to swear more.[25] Researchers randomly assigned participants to play one of two video games, a golf game or a first-person shooter (a first-person shooter involves navigating through a virtual world and trying to kill people with weapons like guns, missile launchers, and knives). Participants were then asked to name as many profane words as they could in one minute. The result was that after the (violent, aggressive) first-person shooter, they were able to name significantly more profane words than after the golf game.

The upshot is that although suggestive, the *Pediatrics* study is only lightly so. In the way that if I polled a bunch of schoolchildren about their favorite color and found that the ones who liked red were more likely to also say they played sports frequently, that would suggest that color preferences cause sports participation. So, very, very lightly.

The current state of our knowledge is this. We have no meaningful evidence that profanity causes aggression, just as we have no evidence that it causes changes in emotional responsiveness. Correlational studies don't help. To know for sure, you'd need to run an experiment, which is the best way to ask a question about causation. You overcome the many limitations of correlational studies by manipulating one thing and seeing whether it has an effect on some other thing. If you do your job well (and designing experiments is not easy), then you can conclude whether or not the thing you manipulated caused a change in the thing you measured. The example everyone is familiar with is randomized drug trials. You give people a drug or a placebo (that's what you manipulate), and you measure the effects on their health. Because you've randomly assigned people to the placebo or the drug condition, a meaningful difference between the two groups in terms of health outcomes means that the drug caused a difference in health compared with the placebo. It can't be the reverse because the arrow of time doesn't work that way. And as long as you make sure that the only difference between the groups is whether or not they receive the experimental drug, then you know there aren't any hidden variables. You can conquer correlation through experimentation.

So an experiment about profanity and aggression might go something like this. You take kids and randomly assign them to conditions that expose them to more or less profanity. There are no other differences between the groups. And after sufficient exposure, you measure how aggressively the children in each group act in some social task. Maybe you have a bully take away a child's favorite toy and see how the kid responds, or something like that. And you look to see whether the kids who, through no fault of their own, were randomly assigned to greater profanity exposure act more aggressively as a consequence.

As you may already be thinking, there's an ethical dilemma here, which may be the reason this study hasn't ever been conducted. In order to run this study, there has to be a reason to think that the hypothesis might be true—that increased exposure to profanity will increase aggression. But increased aggressiveness could be harmful to the child and those around him or her. So if the study found the effect it was set up to test for, it would harm half the participants. And those participants are children, who in general you'd prefer not to harm. As a result, the risk would outweigh the potential benefit, making a study like this a nonstarter.

I posed this problem to the students in my lab, and one of them came up with a clever way around this problem.* Because the experiment I described is ethically impossible, suppose instead you took the opposite tack. What if some people were already opting-in to exposure to some profanity-laden media. Say, people who preordered some video game with swearing in it or who decided to watch an R-rated movie. What if you could convince the software company or the movie distributor to randomly send users one of two versions that differ only in the amount of profanity used. One is the normal version, with all the profanity; the other, an expurgated version without profanity. And then after these people have watched the film or played the game, you measure their aggressiveness in some social task. Neat idea. And it's totally ethical in that the people will be exposed to the video game regardless, and if the hypothesis is correct, then the half of them who get the expurgated version will experience lower risk by dint of being in the experiment. It's a win-win (as long as the only hypothesis in play is that more profanity makes people more aggressive).†

The upshot is this: If exposure to profanity increases aggression through increased acceptance and use of profanity, that fact is in principle

---

* It was Tyler Marghetis, for what it's worth.
† The only drawback I can come up with is the following. Suppose exposure to profanity actually decreased aggression. For instance, suppose that people seek out profanity specifically in order to have an outlet to deal with feelings of aggression. And what if shielding them from profanity actually deprives them of this outlet, thereby causing them to find other outlets for their aggression? If there's reason to believe this (and to date there's effectively as much evidence for this position as for the reverse effect), then the study again becomes unethical. Because as soon as it's possible that your manipulation will harm children, there had better be some really remarkable payoff for conducting the study. This imaginary study would not, for instance, cure cancer.

knowable. There's an experiment to do. It's complicated, but it's doable. But the current state of our knowledge is basically that we have none. We don't know—although there's reason to believe it from other findings in language learning—whether increased exposure to profanity increases children's use of profanity. We don't know whether increased use of profanity increases aggression, and there's as much evidence of this as of any other possible relation between these variables: that increased aggression increases profanity use or that increased profanity use decreases aggression. In other words, we have no idea.

$ % !

So let's take a step back. If there's no evidence of profanity causing harm, then why would so many people—including researchers, journalists, parents, and legislators—believe it? Why would the American Academy of Pediatrics issue a press release describing this research? Why would it seem so plausible?

I think two things are going on here. The first has to do with moral thinking. Many people believe profanity is immoral. Bad words are bad. We'll get into why people think that in a later chapter, but I want to focus here on the cognitive consequences of thinking that something is bad. If you believe that something is good, you're more likely to believe other good things about it, and if you think something is bad, you're more likely to believe other bad things about it. The positive version of this, known as the halo effect,[26] has been studied for a century.[27] The negative version, the horns effect (halo: angel; horns: devil), appears to apply to what you think about everything from people to products. Politicians use the halo effect; if they dress well, sound smart and approachable, and look friendly, then you're likely to think they're good at their job.[28] Marketers use it too.[29] People who like Apple Computer, Inc., for instance, are more likely to think good things about its individual products—that they're high in quality or useful or well designed—than the products of a company they don't know anything about. And something similar is probably going on with profanity. People think *shit* is a bad word. So they're likely to believe that it can also be dangerous, cause harm, and have other negative attributes— more so than they'd believe these things for a "good" word, like *please* or *thank you*. The bar to convince people that *sir* or *ma'am* increases aggression would be far higher than a single correlational, self-reporting study.

People might also be prone to believe bad things about profanity due to a second cognitive consequence of negative beliefs: rationalization. When people have unfounded moral beliefs, they often come up with explanations for them. Those explanations ("rationalizations" or "justifications") may or may not be true, but they do make people feel better about their beliefs. If you believe something is bad, like profanity, but don't really have a good reason why, then you might be strongly attracted to explanations that would justify your belief. If profanity increases aggression, that justifies your belief that profanity is bad. This sort of reasoning from back to front leads people to overlook logical weaknesses in an argument as long as it supports something they already believe.

People who believe that profanity is bad are psychologically biased to believe a variety of other bad things about it, like that it increases aggression or numbs normal emotional responses. In essence, people who believe profanity is bad conclude that little minds are fragile things, perverted and deformed by the linguistic deviants around them. But this gives children too little credit. While the effects of abuse are real, there's every reason to believe that children's minds are resilient to profanity. The real fragility resides in the minds of those adults who are easily swayed to believe in the deleterious effects of profanity.

The $100,000 Word

Kobe Bryant was the starting shooting guard for the Los Angeles Lakers from the time he was drafted straight out of high school in 1996 until he retired in 2016. Over those twenty years, he saw most everything the National Basketball Association (NBA) has to offer. Yet even he was still prone to emotional swings on the court. In a game against the San Antonio Spurs on April 12, 2011, Bryant was whistled for an offensive foul and then a technical foul. He demonstrably disagreed with the calls—as he came off the court, he punched a chair and threw a towel onto the ground. The producers of the live broadcast kept the cameras trained on the commotion, a fact he might not have been aware of when, from the bench, he yelled at referee Bennie Adams, "Bennie! Fucking faggot!" Analyst Steve Kerr immediately suggested to the producer on air, "You might want to take the camera off him right now, for the children watching from home."[1]

Even for a twenty-year veteran like Bryant, sports are emotional. Beyond the familiar thrill and agony of, respectively, victory and defeat, athletes experience anxiety about performance, pain from injury, anger at officials' decisions, relief after prevailing in a must-win situation, and so on. As we've already seen, profanity is a privileged conduit to emotion. Add to the mix the intensity, pressure, competition, and spontaneity of sports, and it's really no surprise that athletes swear.

And they swear a lot. This probably wouldn't stir up too much controversy if we were just dealing with enthusiastic *fuck yeahs* of victory and saturnine *aw fucks* of defeat. But we're not. Oftentimes, as in Kobe

Bryant's case, the profanity belongs to the linguistic third rail of slurs. *Fucking* might not be hard to play off, but *faggot* is going to keep your publicist busy.

Now, to close out this particular episode, Bryant later apologized and explained that he had spoken "out of frustration during the game, period" and that "the words expressed do not reflect my feelings toward the gay and lesbian communities and were not meant to offend anyone." And to his credit, he's refashioned himself as an advocate for LGBTQ rights and respectful speech. For example, when Jason Collins came out as the NBA's first openly gay player, Bryant had this to say:

> I think him coming out was really brave. As his peers we have to support
> him, just rally around him and hopefully everybody else comes out and
> be themselves in who they are.[2]

And when a pair of Lakers fans used a slur on Twitter, he called them out (in an economical 131 characters):

> Just letting you know @PacSmoove @pookeo9 that using "your gay" as a
> way to put someone down ain't ok! #notcool delete that out ur vocab

But his contrition was a day late and a dollar short. Actually, to be precise, it was a tad more than a dollar short. The NBA decided the day after the incident on the court to levy a fine for use of "offensive and inexcusable" words. The bill they sent Bryant: $100,000.

Bryant is just one entry in a long list of basketball players punished for using slurs. In 2012, New York Knicks forward Amar'e Stoudemire was fined $50,000 for a tweet calling a fan a *fag*. In 2013 Roy Hibbert was fined $75,000 for using the expression *no homo* in a press conference. In 2015, Rajon Rondo was suspended for a game for calling a referee *faggot*. And so on. Simultaneously, the National Football League (NFL) has started penalizing players for slurs. The league's officiating video, distributed to teams before the 2014–2015 season, stated, "The NFL will have 'zero tolerance' this season for players' on-field use of racial slurs or abusive language relating to sexual orientation."[3] The penalty? Fifteen yards for the first infraction, as well as possible fines or other disciplinary action. San Francisco quarterback Colin Kaepernick was caught in the dragnet during the second game of that initial season, when he was fined $11,000 for ostensibly using the word *nigger* on the field.*

On the surface, leagues and teams champion these policies as ways to ensure sportsmanship and to create safe and respectful environments. A more cynical interpretation sees them bowing to the same pressures as other entertainment organizations, like broadcast television and the film industry, blanching the entertainment product to the point where no viewer can find anything in it to object to. Regardless of the impetus, the consequences to players' pocketbooks are real, and the message is clear: no slurs tolerated here. And for all of sports leagues' documented evils (baseball's history of segregation,[4] football's suppression of head trauma research,[5] basketball and football's exploitation of unpaid "amateur" college athletes),[6] the antislur campaign in play in the major North American sports leagues is at least one thing that no one could possibly object to.

Even if we accept that profanity in the aggregate doesn't cause harm to children or adults, slurs seem like a different creature. They're built to hurt. In this chapter, we'll see how. But that doesn't mean that banning them is the most productive approach. I'll make the case below that even though slurs may cause harm, blanket policies outlawing them—like those adopted by the NBA and NFL and many other private and public organizations—can actually do more harm than good.

$ % !

Slur-banning policies may at first blush seem reasonable for compelling psychological and historical reasons. To begin with, people find slurs more offensive than any other class of words, including other profanity. We can see this clearly from Kristin Janschewitz's data, discussed in Chapter 1.[7] You might recall that people were asked a number of questions about taboo and nontaboo words. I categorized the taboo words according to the now familiar Holy, Fucking, Shit, Nigger Principle. Naturally, in doing so, I had to make some judgment calls. Is *whore* a slur or a sex-related word? (I classified it as a slur.) Does *jism* pertain to bodily effluvia or to sex? (I went with the latter.) But since these groupings are meant to give a general idea

---

* The league eventually reduced the fine by half due to an absence of evidence that Kaepernick had actually uttered the slur: Sandritter, M. (October 15, 2014).

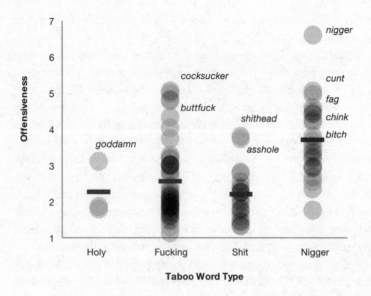

Each dot represents an English word in one of the four major categories of profanity. Slurs are judged far more offensive than any other group.

of how the language populates its profane vocabulary, let's not get hung up on these questionable cases—they turn out not to make a difference for the big picture issue I want to extract. Look at what happens when you split profanity in this way. In the graph above, each word is a dot, and the mean offensiveness rating for each word is on the *y*-axis. So higher-up words are judged more profane on average.

As you can see, the offensiveness of the various words ranges broadly. There's one single, lonely data point all the way on the top right, perched peerlessly at the zenith of offensiveness. That's *nigger.* But although it's an overall outlier, other slurs behave consistently with it. Directly below it is *cunt,* then *fag, chink,* and so on. The horizontal bars show averages by category, and you can see that on average slurs are judged more offensive than any other group by a full point or more. And this might actually understate how offensive slurs are. Some of the most offensive words in the other categories are arguably slurs—the list of most offensive *Fucking* words, starting at the top, are *cocksucker* and *motherfucker.*

But this demonstration probably just confirms what you already know. People find slurs offensive. For some slurs, this is nothing new: they were built to offend from the outset. It's their reason for being. As far as we can tell, *Ching-Chong* is and always was a term of offense. Same with *wetback*, *sand-nigger*, *camel-fucker*, and so on. And just as there are typical sources from which profanity in general draws, particular semantic pathways lead to slurs.[8] A common one is physical characteristics believed to identify members of these groups. Sometimes these terms identify stereotyped skin color, like *yellow* or *Redskin*. Slurs can also identify body features, like *snip-dick*, *slant*, *slope*, or *thick-lips*. They can identify ways people in a group are believed to sound (like *Ching-Chong*). They can come from animal words, like *coon* or *bitch*, and from stereotypical occupations, activities, clothing, or foods, like *cotton-picker*, *towel-head*, *breeder*, *carpet-muncher*, or *cracker*. Using words associated with stereotypical appearance or behavior is an effective way to dehumanize members of a group.

In other cases, otherwise neutral words have grown into slurs, not necessarily from original intent but due to the social contexts of their use. Even *nigger* wasn't always the linguistic powder keg it now is. Nor were *Chinaman* or *cripple*. In the long history of these words, they've gone through changes, just as our attitudes toward language and toward members of minority groups have evolved. And they've followed a similar trajectory. It's worth expanding on that a bit to see how words evolve into slurs.

Nigger originally derives from the Latin root *nigr-*, which just means "black." Before the 1900s, it was the default way to refer to Americans who would come, through the twentieth and twenty-first centuries, to be referred to as *colored*, then *black* and *African American*. For example, *nigger* shows up in Mark Twain's most lauded contribution to American literature, *The Adventures of Huckleberry Finn*, published in 1884. In case your memory is rusty, narrator Huck Finn recounts his escape from an abusive father in Missouri. Along the way, he befriends an escaped slave, Jim, and together they head for freedom in the North. Over the course of their adventure, the word *nigger* appears more than two hundred times—not as a slur but as a generic reference term.

Of course, even if the word itself was largely neutral in the nineteenth century, the context of its use (both in the book and in real life) was anything but. It accompanied people through centuries of enslavement and subjugation. As a consequence, for many people, it remains tainted by that legacy. As social attitudes changed, by the twentieth century *nigger* had started to gain a strong negative connotation.[9] It soon gave way to *colored* and then *black*, *Afro-American*, and *African American*.

The stories of *Chinaman* and *cripple* are similar. Both originally referred to people neutrally. *Chinaman* was as benign as *Englishman* or *Frenchman*, which are still used without negative connotation. In fact, through much of the nineteenth century, Chinese Americans used it in positive contexts. For instance, in a letter about immigration policy to then California governor John Bigler, a Chinese American San Francisco restaurant owner proudly wrote, "Sir: I am a Chinaman, a republican, and a lover of free institutions."[10] *Chinaman* started to develop a negative connotation only around the turn of the twentieth century. And *cripple*, a noun used at least since the tenth century, yielded its place as the default term as late as the twentieth century, when *handicapped*, *disabled*, and their successors came into preferred use.

Again, just like *nigger*, *Chinaman* and *cripple* were default words for a time, but during that period people of Chinese descent and people with disabilities were in many cases treated as second-class humans. And the connotations that terms pick up over centuries of use are not easily shed. For example, the Fritz Pollard Alliance, an organization that promotes diversity in the NFL, has argued, "Whatever arguments people want to make about the 'N-Word' being benign, it reeks of hatred and oppression, and no matter the generation or the context, it simply cannot be cleansed of its taint."[11] It's easy to find similar objections to *Chinaman* or *cripple*.

The different paths that slurs have taken don't matter quite so much as what the words are now. And as we've seen, they're offensive. So it makes sense that when they show up in public domains, like in professional sports, the response from the media, viewers, and league officials is stronger than for other, less offensive types of profanity. This offensiveness itself might be enough for leagues to implement slur bans. Professional sports survive off of the lifeblood of broadcasting contracts: if viewers are offended by what they see, they'll tune out and take advertisers and lucrative television deals in their wake. And the leagues might not have dug any deeper than this.

But the argument for slur bans is actually even stronger. Although I doubt anyone in the league offices in question has read this research, there's also some evidence, as we'll see in a moment, that exposure to slurs causes psychological and social harm. Now, it turns out that the effects of slurs actually differ depending on whether the person exposed to them is a member of the slurred group or not (if you're heterosexual, for example, the word *faggot* affects you differently than if you're homosexual). But let's wade into the reeds here a bit.

We know the most about how slurs affect people who are not members of the defamed group. Suppose you overhear slurs like *nigger* or *faggot*, and those terms do not refer to you. How does that change your feelings about the people those words refer to? Several studies have examined this question. One conducted in Italy had heterosexual participants perform a free-association task. Presented with a list of words, like the Italian equivalents of *sun*, *American*, and *lion*, participants had to come up with three related words for each. And the key manipulation was that the last prompt word on the list was either *gay* or *faggot* (since the study was conducted with Italian participants, the word wasn't actually *faggot* but the Italian equivalent, *frocio*).[12] After finishing this free-association task, participants performed a totally different task: they made a recommendation about how the city should spend money. They were told that the city council was deciding how to allocate funding to two distinct programs—one working toward AIDS-HIV prevention for "high-risk groups" and the other working on fertility issues in young couples. Their job was to decide how much of a fixed amount of money the city council should dedicate to each. The logic was that if *gay* and *faggot* had different effects on how the participants thought about themselves in relation to homosexual people, then this should affect their decisions to allocate funding toward a program more likely to help homosexuals (AIDS-HIV prevention for high-risk groups) or to one more obviously oriented toward heterosexuals (fertility). And that's what they found. People originally given the word *faggot* to free-associate from were far less likely to allocate resources to the AIDS-HIV prevention program than those who free-associated on *gay*. In other words, exposure to a slur can bias people against sharing resources with members of the defamed group.

Exposure to a slur also affects how nonmembers of the defamed group think about members of the defamed group. One study again presented heterosexuals with neutral or derogatory group labels, *gay* or *faggot*, and then afterward asked them to select from a list words they associated with homosexuals and other words they associated with heterosexuals.[13] The study was again conducted in Italy, and the list of words participants had to choose from included Italian words describing humans (like *person*, *citizen*, and *hand*) and others describing animals (like *animal*, *instinct*, and *paw*). And when the researchers tallied the results, they found that people picked more animal-related terms for the homosexual group and more human-related terms for the heterosexual group, but only after hearing the slur *faggot* and not the neutral group label *gay*. To make sure this was really about a slur directed at that group and not about derogatory terms in general, they included another condition in which people performed the same task after hearing the Italian equivalent of *asshole*, which is *coglione*. Unlike *faggot*, *asshole* did not lead to different apportionment of animal or human terms between the two groups. So this suggests that there's something literally dehumanizing about slurs—something that makes outsiders think about defamed group members as though they have fewer human attributes.

Exposure to slurs even affects how you physically interact with members of the defamed group. A study at the University of Queensland in Australia[14] subliminally presented one of three words to people by flashing it on a computer screen. It appeared forty times but much too briefly for the participants to consciously apprehend it. The three words were *gay*, *faggot*, and *asshole*, and each participant saw just one of them. Then they went into another room for a putative discussion with a student, Mark, about the situation of homosexuals at the university. Mark, they were told, was homosexual. Participants were directed to enter the room to wait for Mark and to prepare for the interview by setting up chairs for the two of them. But the researchers were less interested in the interview than where the participants placed the chairs—more specifically, the distance between them. Participants placed the chairs about four inches farther apart when they had been subliminally primed with *faggot* as compared with *gay* or *asshole*. A slur drove people to keep themselves physically farther away from members of the defamed group.

So there's some evidence that exposure to slurs about others leads to biases against those people—financial, psychological, and physical. But what about the direct effects on the people that the slurs are about? What does *faggot* do to you if you're homosexual? This is a more complicated story. The literature reveals that there's been almost no work on this question. And that's probably for ethical reasons. If you think that exposing people to slurs directed toward their group could cause them harm, then it's hard to justify a study like that unless there's substantial benefit to the participants or to society. There might well be; it's just not the easiest case to make. So we know very little. And what little we do know shows that the effects of slurs on members of the defamed group don't track with their impact on outsiders.

We can see this from another Italian study that used a "lexical decision" task. Lexical decision is a type of experiment in which you have to decide whether a string of letters that appears on a computer screen makes up a word in your language or not.[15] The length of time it takes you to decide in the affirmative reveals the current state of various mental operations. For instance, people are known to respond faster to a word like *dog* when it follows a related word like *puppy*, which tells us that thinking about puppies activates thoughts about dogs and perhaps the word *dog* itself. In the study in question, some of the words participants saw were adjectives that describe culturally relevant, positive perceived aspects of homosexual males (like *elegant* and *artistic*); others described perceived negative aspects (like *effeminate* and *emotional*). The trick was that before each word, a neutral term like *gay* or a slur like *faggot* blinked on the screen so quickly (for only fifteen milliseconds) that the participant would only process it subliminally.

The researchers found that people's speed in deciding whether a string was a word in Italian or not was influenced by whether they had unconsciously been exposed to the neutral term *gay* or the slur *faggot*. But whether the participants identified as homosexual or not determined how they were affected. When heterosexual participants subliminally saw *faggot* rather than *gay*, they were slower to decide that positive attributes of homosexuals like *elegant* and *artistic* were words of their language. This makes sense if you think that slurs, even processed subconsciously, bring up negative attributes and suppress positive attributes of the targeted group. You see *faggot* unconsciously, and then you see a positive attribute of homosexuals, and it takes you a little longer to read and understand that word because it's inconsistent with the subconscious framing of homosexuals induced by *faggot*.

But homosexual participants were affected in a totally different way. Seeing *faggot* rather than *gay* slowed their recognition of the negative attributes, like *effeminate* and *emotional*. Indeed, by comparison, *faggot* made them think faster about positive aspects of their own group. This might seem surprising to you. It was to me when I read it. But here's a possible explanation. When threatened by an external source, some people have a tendency to retrench within their group identity. And perhaps that was happening here. Subliminal exposure to a slur might have gotten homosexual participants' backs up, leading them to feel stronger identification with their defamed group. They would thus evince more positive feelings about themselves and their membership in that particular group. That in turn would lead to faster reactions to positive adjectives describing homosexuals. But this is just one of several possible explanations. For instance, perhaps homosexual Italians have reappropriated the word *frocio* ("fag"), which has even developed a positive connotation. More on that later.

It's important to be clear about what these data do and don't mean. Subliminally presenting slurs to members of defamed groups might lead them to process positive in-group attributes faster. But that doesn't mean that calling people by slurs is good for them. Homosexual people (like heterosexual people) deem slurs highly offensive. Remember that Janschewitz's data has *nigger, fag,* and *cunt* at the very top of the list, and other studies find the same thing.[16]

Moreover, a single word presented in isolation may not have the same impact it can in context. There's some circumstantial evidence that unlike other types of profanity, slurs, when deployed as part of bullying, may actually cause harm.[17] A study in the *Journal of Early Adolescence* asked middle schoolers about their mental health and school experience twice—in seventh grade and then again in eighth grade. Researchers were interested in whether being called by slurs during that year correlated with changes in students' well-being. So in eighth grade, they also asked the children how often they were called by homosexual slurs, such as *homo, gay,* and *lesbo.* The researchers found that the boys (interestingly, not the girls) who reported being subjected more frequently to homosexual slurs were also more likely to exhibit an increase between seventh and eighth grade in anxiety, depression, and personal distress, as well as a decrease in their sense of school belonging. We have to be careful how we interpret this correlational study (see the preceding chapter!), but at the very least, this result doesn't suggest that slurs are good for children.

Finally, it's possible that slurs could negatively affect in-group members in another way, known in the social psychology literature as "stereotype threat." Members of particular groups are socially stereotyped as bad at certain things—for instance, there exist in the United States stereotypes about females and certain ethnic minorities being less proficient at math and science than their male or Caucasian peers.[18] And in fact, educational psychologists have observed that members of these stereotyped groups do in fact perform worse on average in those particular areas, but only under certain conditions. When exposed beforehand to negative stereotype information about their gender or ethnic group—for instance, after being reminded that "women do not perform as well on this test as men do"— they perform significantly worse than when they hear positive information about their social group, like "women tend to be creative, and success on this exam depends on creativity."[19] While I don't know of any direct evidence addressing the question, it's reasonable to hypothesize that hearing a slur directed at you (*bitch, nigger, wetback,* and so on) creates a threatening environment that leads to poorer performance in those enterprises that your group is stereotyped as bad at.

So in a nutshell, not only are slurs judged offensive, but they also have demonstrable negative effects on how outsiders treat members of defamed groups. It's not as clear that just hearing the words has direct negative effects on members of those same groups, even if it seems likely that they would. That's more complicated.

And so, a preponderance of evidence shows that slurs offend people, evoke dehumanizing and discriminatory behavior toward members of defamed groups, and either were designed or evolved to insult and oppress. Considering these facts, it makes sense that people would call for an end to such words' use. The Fritz Pollard Alliance writes, "While we understand and respect that different generations have different means of communicating, we cannot condone on any level the use of the 'N' word. . . . Simply put, from this day forward please choose to not use the 'N' word. Period!"[20] The National Association for the Advancement of Colored People, in a similar call, even held an elaborate funeral ceremony for the word in Detroit in 2007. A coffin with the word *nigger* printed on it was buried. Detroit mayor Kwame Kilpatrick, who was in attendance, declared, "Good riddance. Die, N-word. We don't want to see you around here no more."[21]

With all the trouble that words like *nigger* and *faggot* cause, fining a millionaire athlete for using them probably doesn't seem like the worst decision in the world.

$ % !

But here's the problem. People don't use these words in just one way. Yes, the terms serve to defame and insult, but they have other lives as well. There's been more linguistic analysis of *nigger* than any other English slur, so let's focus on it. The first clue that the word has different manifestations comes from the fact that it's spelled variably: as either *nigger* or *nigga*. What's the difference? There's no question that *nigger* is a slur. But when it's spelled *nigga*, a certain group of people use it differently. These people are mostly native speakers of a particular variety of English often called African American English (AAE). Speakers of this variety of English, most but not all of whom are African American, have adopted the word and run with it in ways that have largely (but arguably not entirely) liberated it from its defamatory connotations. As rapper Tupac Shakur explained to MTV reporter Tabitha Soren, "*Niggers* was the ones on the rope, hanging off the thing; *niggas* is the ones with gold ropes, hanging out at clubs."[22] Not every speaker of AAE would agree with this characterization, but it reflects the sense that there are two words here—one used to defame and suppress and another that's an in-group term of positive self-identification.

It's counterintuitive that the very people most injured by a defamatory word would adopt it as their own. But this same reappropriation of slurs has happened over and over again across the world and across denigrated groups. Instead of using offense at certain derogatory group labels as an impetus to try to decrease use of those terms, some people co-opt and re-cast them as badges of proud group membership. Some African Americans use *nigga*, just as some homosexuals use *queer* or *faggot*, and some women use *bitch* or *slut*. These reappropriators often feel they can mitigate a word's power to offend or hurt by taking ownership of it.

There is a little evidence that this works. A 2013 research paper reported on ten studies designed to understand the ecology of slur reappro-priation—who does it, when, and what effects it has.[23] The most revealing experiments explored participants' responses when someone else used a slur to describe him- or herself. Researchers randomly assigned people to read a newspaper article in which someone belonging to a stigmatized group was described with a slur, like *queer* in one experiment and *bitch* in another. In each experiment there were two conditions, which differed only in terms of who used the derogatory term. For half of the participants, an outsider was quoted as saying, "You're queer" or "Your name is bitch," to another individual. For the other half, the homosexual or female person in question was quoted as saying, respectively, "I'm queer" or "My name is bitch." After reading their assigned version of the article, participants rated how negative the given word was, on a scale from 1 to 7. For *queer*, the difference was massive. Participants who read, "You're queer," rated *queer* as extremely offensive, with an average score of nearly 7 out of 7. But those who read, "I'm queer," rated it about 4.5 in offensiveness. The *bitch* exper-iment produced the same type of effect, in the same direction, but wasn't quite as pronounced. Ratings swung from 5.9 following other-labeling of *bitch* to 4.7 after self-labeling—still a significant difference.

The upshot is that when you observe people using a slur to describe their own group, that word seems less offensive to you, at least in the short term. But we don't have any experimental evidence about potential longer-term effects of self-labeling on feelings about the slur—either among the people who self-label with it or outsiders.

We do know that once reappropriated, a slur expands and morphs into something nearly unrecognizable, at least linguistically. *Nigger* has changed a lot since it became *nigga*. Some really revealing data comes from a paper presented at the Linguistic Society of America conference in 2015. The researchers, who speak AAE, documented the use of *nigga* in contemporary English by analyzing 20,000 tweets containing the word.[24] They found that *nigga* retains very little similarity to *nigger*.

For one thing—and I think many people will be familiar with this use—*nigga* serves as a generic noun that neutrally refers to usually male, usually African American people. I say "usually" because each of these generalizations has rampant counterexamples.[25] For instance, it can occasionally refer to nonhumans. An attested example from the study is this tweet: "I adopted a cat and I love that nigga like a person." It can also refer to people who are not African American, as in this tweet: "This white nigga just slapped his mom x_X." So the word has migrated substantially. This generic, mostly human, mostly male use of *nigga* is roughly equivalent for many speakers of AAE to other generic terms, like *guy* or *dude*.

But here's something you probably didn't know. *Nigga* can actually behave like a pronoun too. As a refresher, pronouns include words like *I, you, we, she,* and *they* that can stand in for specific nouns. To a linguist, this is a huge deal. Languages don't just add new pronouns willy-nilly. Just ask people militating for gender-neutral pronouns in English. Never heard of *ze* and *zir*? I'm not surprised. Pronouns are part of the grammatical core of a language and are appropriately resistant to change. But according to how it's used on Twitter, *nigga* has crashed the pronoun party. Or more accurately, *a nigga* has. Consider examples like this one, again from a real tweet: "Spring got a nigga feelin myself." We have to put on our grammarian hats to know that this is a special use of *a nigga*. Who does *a nigga* refer to? The hint comes from *myself*, which you'll remember from Chapter 6 is a reflexive *-self* pronoun that tells us that *a nigga* absolutely has to be referring to *me*. So in this use *a nigga* appears to be acting like the first-person pronoun *me*. It can also act like *I*, as in the tweet "A nigga proud of myself."

Or it can be like *my*, as in "You read all a nigga's tweets but you still don't know me." In all these cases, *a nigga* appears to be acting like a first person pronoun. It refers to the speaker.

To reiterate, these ways of using the word—as a generic noun or a pronoun—really only appear in certain varieties of English. And African Americans or people who culturally identify with African Americans happen to mostly speak those varieties. That fact will become important in a moment, so hang on to it. For now, let me just reiterate that *nigger*, having been appropriated by the very people it's meant to denigrate, has assembled new uses that may be totally unknown to people who speak other varieties of English,[26] like most national legislators or the owners of most professional American sports teams. The same process of reappropriation followed by changes in meaning and grammar has happened in the histories of *faggot*, *slut*, and other co-opted slurs. That's point number one in the argument that banning slurs can be counterproductive.

The second point is that context matters. In certain contexts it's socially acceptable to use slurs even if you don't belong to the group they denigrate. One such context is in tightly knit groups of young people. It's been most studied in groups of males but is also apparently present among females.[27] As I was once a young person myself, I can personally attest that it's not uncommon for the one Jew in the group (for instance) to be razzed as a *kike*, *hebe*, *big-nose*, *Jesus-killer*, *snip-dick*, and so on. A homosexual in the group might be addressed as *fag*, *butt-dart*, *ass-spelunker*, *pillow-biter*, or *shit-pusher*. Although in other contexts these words might elicit offense, among groups of close peers, a different dynamic is at play. On the insulter's side, using a derogatory term highlights how close he is to the person he's insulting. If he can get away with using a slur, even though he's not a member of the group insulted by it, that shows that he must be very close in another way. At the same time, on the insultee's side, allowing others status as an "honorary" Jew, homosexual, or whatever and permitting them to use slurs that would usually offend demonstrates how secure he is in the friendship.

When we're talking about groups of young people, posturing is often also involved, and allowing others to insult you gives you two ways to exert your dominance. First, by allowing insults, you show your own self-confidence—you demonstrate that mere words don't bother you. And second, you have the opportunity to engage in one-upmanship via verbal sparring. This can take place in an impromptu manner. Or it can be part of a ritual insult game, like "the dozens"—there are other names for it, like "snaps" or "signifying"—that's performed mostly, but not exclusively, by young adult males, often in working-class or poor neighborhoods. Basically, participants spar verbally, insulting each other and the people and things the opponent values, like his relatives, especially close female relatives. The point is to do so in as creative, insulting, and specific a way as possible. For instance, a person playing the dozens might say *Yo mama's so fat, when she was diagnosed with a flesh-eating disease, the doctor gave her five years to live.** Verbal sparring is often filled with slurs—but again, these are licensed by the social environment. The goal is to simultaneously display verbal agility and a superior ability to stay cool under fire.†

Within these particular communities of practice, slurs operate not as insults but as part of a socially licensed interaction. It's similar to how actions taken on the battlefield (like shooting people) or, for that matter, the football field (tackling people) are socially permissible in those contexts but not elsewhere. And as a result, in some contexts slurs aren't intended to offend and are not received as offensive. They can be poetic, creative, and even important to creating and reinforcing social relations.

So the upshot is this: Some slurs are used as much, if not more, by members of the groups they originally denigrated. These in-group members use the terms in ways largely divorced from their original negative connotation. In some contexts the use of slurs even reveals and reinforces group coherence and personal allegiances.

This brings us back to the issue of banning slurs and the consequences of doing so.

$ % !

Many organizations representing the rights of specific groups advocate against anyone ever using terms perceived as pejorative. They reason, broadly, that using derogatory group labels can not only cause social tension but also disempower members of less powerful groups. This sentiment,

---

* There appears to be room at the bottom of this page for more *yo mama* jokes. You know, for the sake of science:
    Yo mama's so fat, she went to the zoo and the elephants started throwing her peanuts.
    Yo mama's so fat, her ass has its own congressman.
    Yo mama's so fat, she's got smaller fat women orbiting around her.
    Yo mama's so fat, on Halloween she says, "Trick or meatloaf!"
† Rap battles often have this same format, though in a more structured environment, and the practice has an early antecedent in the ancient practice of flyting—as in Conlee, J. (2004).

translated into sports league policy, justifies punishments for using slurs—
$100,000 for *faggot*, fifteen yards for *nigger*.

But even if well-meaning, a policy that legislates words, without taking
into account how they're used or by whom, runs the risk of causing dis-
proportionate and unfair injury to the very people it aims to protect in the
first place. This is most obvious with *nigger* because it's used so much more
frequently by African Americans and because it manifests in so many ways
other than as a slur. As former NBA star and current analyst Charles Barkley put it, "I'm a black man. . . . I use the N-word. I will continue to use the
N-word among my black friends and my white friends."[28] Many US athletes
are African American (76 percent of NBA players and 66 percent of NFL
players),[29] which means that many of them are also native speakers of AAE,
the language variety in which *nigga*, as we've seen, acts as a common, inoffensive noun as well as a pronoun.

The consequence is clear. To ban *nigger* is to disproportionately silence
and punish the very people the regulation ostensibly strives to protect.
We're going to protect you, the league office is saying, by cutting out part of
your language. And if you use this particular first person pronoun, you're
going to be playing for free on Sunday, if we allow you to play at all.

I'm not the first person to observe that this policy is the linguistic
equivalent of a frontal lobotomy. When asked about a possible ban on
nigger, Stanford-educated Seattle Seahawks cornerback Richard Sherman
said, "It's an atrocious idea. . . . It's almost racist to me."[30] He pointed out,
"It's in the locker room and on the field at all times." And "I hear it almost
every series out there on the field." Sherman speaks AAE. But if you were a
league executive or a team owner who wasn't a native speaker of that variety of English, this might not occur to you.

Now, to be clear, this is not a First Amendment issue. Sports leagues
are private companies and can ban whatever they like. So the fact that the
law views slurs in general as protected speech doesn't matter—an employer
can set its own policies. But I'm not making a legal argument here. I'm
saying that if someone thinks banning words is a silver bullet that will
eradicate racism, sexism, heterosexism, or any other offensive ism, with no
downside, he or she is mistaken.

Nor am I arguing that the right policy approach is to legislate intent.
No one wants to have to infer what someone really meant by a particular
word—not referees in the moment or league executives after the fact. Was
that *nigger* or *nigga*? Was it a slur or a pronoun? Do these two players like

each other enough to legitimately use the term in a socially licensed way? Intent is essential in the courtroom, where it has to be established to determine whether a defendant has committed a crime. But intent is really hard to infer even in the legal setting and even with the full power of subpoenas, sworn testimony, and lengthy reflection. There's no reason to think it would be feasible to impose a courtroom standard on the basketball court or that doing so would lead to fair or reasonable outcomes.

So what's the most productive response? To do nothing? Let's zoom out. I've reviewed some evidence that slurs are offensive and—unlike other kinds of profanity—can even plausibly do harm in certain contexts. But those same words have very different uses in other contexts, some of them positive, and sometimes among members of the very groups ostensibly harmed by those words in the first place. As a consequence, blanket bans or attempts to infer intent are probably more harmful than doing nothing at all to regulate language. I've made this argument with respect to sports leagues, but the same logic applies wherever similar conditions are met: in businesses, schools, or public spaces. The power that slurs have over our brains and bodies compels us to act. But reactive regulation isn't the answer. The next chapter explores some alternatives.

11

The Paradox of Profanity

Profanity is powerful. Its repercussions can be measured in your body. A single *fuck* or *nigger* hastens your heart rate and opens the pores of your palms. Its impact can also be gleaned from your behavior. *Fag* makes you scoot your chair away from someone you think is homosexual and makes you think of him as less human. A raised middle finger leads you to interpret people's actions as more aggressive.

And many of us treat profanity as not just powerful but bad. It's not uncommon to encounter the belief that it betokens an uncreative or lazy mind or a weak vocabulary.[1] The fact that these bad things remain unproven provides little shelter to the offending words.[2]

These negative beliefs that people have about taboo words and their power often lead to attempts at suppression, not just of slurs, as discussed in the last chapter, but of profanity in general. Self-censorship occurs inside your own head when you self-monitor—internally tracking what's likely to come out of your mouth next and stepping on the verbal brakes when something taboo is in the works. You also censor interpersonally when you suppress the use of profanity by other people, especially children. For example, you might react to a child's utterance of a particular word by explaining that it's not acceptable or appropriate—that it's a "bad word." You might go farther and chide or even punish him or her. There's a long history of punishing children for profanity both verbally and physically, washing out their mouths with soap being one of the most creative and most memorable to its victims.[3]

And we also display a suppressive reaction to profanity as a society through social and legal institutions. The sports leagues from the last chapter provide a glimpse into how this works on a small scale. In the United States, the three biggest players are the Motion Picture Association of America (MPAA) and the Entertainment Software Ratings Board (ESRB), which regulate the content of films and video games, respectively, as well as the Federal Communications Commission (FCC), a state organ that regulates broadcasts over the public airwaves. As I briefly mentioned in Chapter 1, one of these institutions' most visible functions is to suppress profanity.

And yet, all the evidence suggests that existing efforts to squelch profanity are ineffectual. Below, I'll present the case—not just for slurs but for "bad words" in general—that there are better ways to deal with profanity than to suppress it.

$ % !

Let's begin with censorship at the societal level and look at how it works and why it doesn't.

The MPAA is an industry organization, founded by the motion picture industry. Among other things, it's responsible for the film ratings that limit children's access to movies.[4] In addition to violence, drug use, and sex, the MPAA identifies "strong language" in films as inappropriate for children of certain ages. Despite its regulatory role, the MPAA has no published standards for what language leads to what rating—no list of offending words or accounting of how many times each word is permitted to merit what rating. The association's method is largely opaque. Production companies submit their films to the board before distribution, and the MPAA ratings board, whose members are an anonymous "independent group of parents," issues a provisional rating. The net result is effectively a sort of censorship. Filmmakers and production companies often self-censor their films in order to reach the largest audience possible, and the ratings board acts as a gatekeeper that can issue seemingly arbitrary requirements and restrict access to the film.

The ESRB acts effectively like the MPAA, but its purview is video games.[5] Like the MPAA, it uses unpublished criteria for determining profanity, but by observing the ratings it assigns, we can infer that they depend on the frequency and intensity of strong language. One study that tried to reverse engineer the ESRB's criteria found that profanity is almost always absent from games rated E (for everyone) or E10+ but is present in 34 percent of games rated T (teen) and 74 percent of those rated M (mature).[6]

The FCC works quite differently. As a federal commission, it's legally empowered to oversee all transmissions over public airwaves, which includes broadcast television and radio. The FCC has charged itself with, among other things, enforcing laws that prohibit profanity during daytime and evening hours. Like the MPAA and the ESRB, the FCC has no published list of banned words, but it describes profanity as "language so grossly offensive to members of the public who actually hear it as to amount to a nuisance."[7] Punishments it can dole out range from issuing a warning to imposing fines and even revoking a station's broadcast license. This has a colossal impact on the language used in music and television. Television production companies and music labels self-censor, avoiding language they think will incur a rebuke from the FCC or from television and radio stations. They or the stations excise profane words that do make it into the artistic product. This happens in a variety of ways, including bleeping offending terms, making alternate versions that replace taboo terms with others, and silencing profanities out entirely, among a number of others.*

There are plenty of potential objections to what the FCC does, especially the mysterious way it defines profanity. But according to the US Supreme Court, the commission does have the right to function as it does. We know this from a 1973 decision, in which the FCC squared off against a radio station that played George Carlin's profane standup routine "The Seven Words You Can't Say on Television."[8] The routine is (ironically) about the words that the FCC bans, which Carlin identifies as *shit, piss, fuck, cunt, cocksucker, motherfucker,* and *tits* (although, of course, there's no official list). He then goes on to elaborate on his impressions of each word. (According to Carlin, *tits* sounds friendly, while *fuck* has a little something for everyone.) The FCC reprimanded the station, which challenged the FCC's right to censor profanity as a violation of the right to free speech. But the Supreme Court upheld the FCC's action, stating that the government has a compelling interest in "1) shielding children from patently offensive material, and 2) ensuring that unwanted speech does not enter one's home." It also ruled that the FCC has authority to prohibit "indecent" broadcasts

---

* A whole host of censorship strategies has been deployed, ranging from repeating the word prior to the profanity to superimposing a sound other than a bleep. One of the cleverest I know of is in Missy Elliot's song "Work It," which uses an elephant trumpet to replace a noun in the following couplet:

   If you got a big [elephant trumpet], let me search it;
   And find out how hard I gotta work ya.

There appears to be no unbleeped (or rather untrumpeted) version of the song.

during daytime hours (when children might encounter them). And most importantly, it decided that the FCC can determine for itself what constitutes indecency.

In addition to these media regulation bodies, the legal system often acts as a de facto censor of profanity. As we saw in the case of the mom arrested for saying *fucking bread* around her kids in a supermarket, laws exist all over the country that, interpreted in specific ways, make it illegal to use profanity, especially around children. There's a good case to be made that these laws violate our First Amendment free speech rights:[9] as we've seen, there's no evidence that profanity of the Fucking type intrinsically causes harm. But logic hasn't often prevailed in this arena. The bans exist as a reaction to profanity, with the intent of negatively sanctioning people who deviate from normative linguistic behavior.*

But these efforts have been largely for naught. People are still being exposed to profanity, and they're still using it. Exposure to profanity on television has not decreased over the last three decades; if anything, the frequency of strong profanities has increased.[10] As musician and professor Tom Lehrer put it, "When I was in college, there were certain words you could not say in front of a girl. . . . Now you can say them, but you can't say *girl.*"[11] Films and video games similarly have seen no decrease in the use of profanity in a similar time frame, with profanity increasing especially in video games. The type of video game that now receives a T or M rating, either of which overwhelmingly includes profanity, essentially didn't exist twenty years ago.

Nor is censorship through bleeping or other local strategies effective. People still infer what the bleeped words were. One clever study[12] had people read sentences with either profane words (like *This custard tastes like shit*) or censored versions (like *This custard tastes like s#!t*). After reading the sentences, participants performed a memory task to see whether they remembered exactly what they had seen. They would see one of the two sentences, with *s#!t* or *shit*, and had to say if this was the exact sentence they saw before. When *shit* replaced *s#!t*, most people had no idea. A full

---

* Even words that merely sound like profanity are often tainted. For example, in 1999, the assistant to the mayor of Washington, DC, David Howard, was forced to resign his post for calling a budget *niggardly*. So was a Florida drug counselor who told a client that he was being *niggardly* about opening up during his drug rehabilitation. To be clear, *niggardly* has nothing to do with *nigger*, either in its history (it's related to *niggling*) or its current meaning (it means "stingy"). And yet, the mere similarity with a taboo word is enough for a repressive reaction to kick in. See Dowd, M. (January 31, 1999) or Mayo, M. (November 11, 2011).

59 percent of the time, people answered that they had seen exactly that sentence previously. In other words, more than half of participants had encoded the word *shit* in memory and thought they had seen the uncensored word, even though it was actually censored to begin with.

Bleeping also doesn't decrease people's impression of how much profanity the program they're watching or hearing contains. One of the few studies to investigate the effects of bleeping presented people with one of two different versions of *A Season on the Brink*, an ESPN biographical documentary about basketball coach Bobby Knight.[13] The uncensored version included seventy-six instances of curse words—Knight is renowned for his vitriolic temper and quick trigger with colorful language. The censored version had all those words bleeped out. People were shown one version or the other, then asked to rate how offended they were by the program and to estimate how much profanity it used. People thought the bleeped version was significantly less offensive, as you might expect. But surprisingly they also thought that the bleeped version had more profanity: people who saw the bleeped version estimated that there were fifteen more curse words in the film on average than people who saw the unbleeped version. So bleeping increases the perceived frequency of cursing.

Even when censorship does have the desired effect of limiting exposure to certain terms, it has unintended and counterproductive consequences. Language is a moving target, and outlawing a word today will inevitably lead to new words sprouting up in its place tomorrow. Censorship is like a game of linguistic whack-a-mole. For example, when Matt Stone and Trey Parker made their feature-length *South Park* film, they were aiming for an R rating. But the first cut they submitted to the MPAA ratings board came back rated NC-17, which would have reduced its potential viewership to a small fraction and killed its chances of breaking even financially. So the filmmakers submitted a series of revisions, responding to each edit the ratings board recommended. But because Stone and Parker aren't exactly fans of censorship (their movie itself is an anticensorship screed), they tried to get away with every bit of vulgarity they could. As Matt Stone said to a *Los Angeles Times* reporter, "If there was something they said couldn't stay in the movie we'd make it ten times worse and five times as long. And they'd come back and say 'OK, that's better'."[14] For instance, it's been reported that the film originally included a song called *Motherfucker*, but that when it was rejected, Stone and Parker replaced it with the arguably equally offensive (and certainly more novel and memorable) *Unclefucka*. Additionally, sources report that the title was originally *South Park: All Hell Breaks Loose*, but the MPAA categorically rejected the word *Hell*, so the film was retitled with a plausibly more offensive double entendre: *South Park: Bigger, Longer and Uncut.*[15]

In sum, the various ways we react to profanity by trying to limit it are grossly ineffectual. They generally don't decrease how much it's used, and even when they do, new words spring up in their place. Bleeping or other word-internal censoring strategies still activate the same words in the listener's or the reader's mind to the point where they're usually indistinguishable in memory from the real thing.

$ % !

But trying to ban language is more than just ineffectual. The practice is actually its own worst enemy. Here's what I mean.

We know that taboo words aren't taboo because they're intrinsically bad. We've seen over the course of this book that profane words are just words; they're made up of sounds and enter into similar (although not always identical) grammatical patterns to other words. There isn't a fixed set of profanity in a language—words meander into and out of taboo-ness. Over time, words move fluidly from banal to profane and back again—think about the histories of *cock* and *swive* (the now deceased, archaic predecessor to *fuck*). Nor is there anything unique or defining about what taboo words mean: even if they tend to draw from certain semantic domains, they can denote the very same things as mundane words (like *penis* and *copulate*). And in fact, a culture doesn't even have to have taboo words if historical vicissitudes haven't conspired to give it any. In other words, there's nothing deterministic about any particular words having to be profane in any given language at any specific time.

And that means that our beliefs about profanity are largely a social construct. The same word can provoke radically different reactions in different cultures or at different times. In Great Britain the word *wanker* is bad, slightly worse than *nigger*. On this side of the pond, it doesn't even register as profane. And even within a country, when people speak different varieties of the language, there's remarkable variation. *Nigger* is profane, except, as we saw, among some speakers of African American English, where *nigga* is a commonplace word that can be used positively and pro-socially. Profanity isn't fixed. It's variable, it's context-sensitive, and it's relative. It's the product of cultural attitudes toward specific words, attitudes that can differ radically from person to person and from culture to culture.

But for these cultural beliefs to exist, they must somehow be instilled. They must be propagated. And this is where it gets interesting. How do you know which words are the bad ones? Think about your own life experience. How do you know that *cunt* and *nigger* are bad?

I suspect you'll come to the same conclusion as sociologists. The things that create and perpetuate these normative beliefs about words stem from the contexts in which those words are used. You can infer what people mean to do with a word by observing how they use it. If you see someone acting violently toward another person while using a slur, such as shouting *You're a fucking cunt!*, that's good evidence that the word is meant to cause harm. This is surely part of the story. But notice that the situations in which people use the words aren't enough. For instance, you wouldn't infer from *You're a fucking cunt!* that all the other words in the sentence, like *you* and *a*, are also meant to cause harm. So in addition to learning about words from how people use them, you also learn from how they avoid them.

Profane words like *fucking* and *cunt* are socially suppressed. People have told you that they're bad. As a child, you might have been scolded or spanked. When your uncle stubbed his toe and yelled *holy fucking cock ass fuck!*, your mother might have chastised him for swearing in front of the kids. There are subtler signals too. When parts of words are bleeped out, or even when you simply notice that people use words in informal settings that they avoid in more formal settings, you learn that those words are not socially appropriate. Adults act as though some words aren't to be said in public, by children, or around children. Children learn precisely that lesson. That's where taboos come from. That's what makes those words profane.

In other words, paradoxically, the taboo words in a language are taboo because of the very actions people take to limit their use. The remedy is the cause.

Profanity isn't special in this regard—cultural norms in general are regularly propagated from person to person and from generation to generation via personal behavior and social and legal institutions that constrain taboo behaviors in certain contexts. I'm sure you can think of many, many socially constructed taboos that go hand in hand with personal and institutional opprobrium. Polygamy, for instance, is commonly practiced throughout parts of northern Africa, the Middle East, and Southeast Asia and the Pacific.[16] But it's taboo in many countries, especially countries with histories of important Christian influence, like the United States, where it's legally banned. Other taboos that vary cross-culturally and are accompanied by personally or institutionally imposed sanctions include things like incest (legal in Côte d'Ivoire, Spain, and many other countries)[17] and open defecation (practiced by 1 billion people worldwide).[18] In places like the United States, not only laws but early childhood instruction, admonitions, and punishment by caregivers and authorities reinforce taboos about these behaviors.

But compared to polygamy, incest, and open defecation, which could arguably compromise the health of communities that practice them, most profane words (perhaps except slurs) pose little known danger. Mores about taboo words in general are more similar to stylistic faux pas. For example, wearing a mustache has been seen as dirty and suspicious in different times and places in the United States. I'd guess that wearing Bermuda shorts to a funeral would meet with disapproval most anywhere in the country. Facial tattoos and piercings have historically been banned or otherwise negatively sanctioned—and still are to varying extents. These behaviors violate social conventions not because they endanger public health but because of a tacit social agreement about the things we do and do not do. But there's nothing intrinsically bad about Bermuda shorts or mustaches; the taboos reflect the cultural values we apply to them. Likewise, the actions people take to negatively sanction profanity create the norms surrounding it.*

So by actively prohibiting profanity, we're acting like a dog licking its wounds. Let me spell out the analogy. Dogs (like humans) have a natural inclination to lick injuries, and for good reason: saliva may speed healing. But there's also a downside. Overlicking can lead to a granuloma, a lesion potentially infected with staphylococcus. A suppressive response can turn a tiny nick into a large, enflamed, and infected hot spot.[19] Like dogs' wound licking, our response to profanity creates a runaway process because it exacerbates the conditions that trigger a progressively more and more aggressive response.

We shouldn't be surprised to see these normative behaviors in ourselves and our neighbors. Many of us have been inculcated with these beliefs, just as with other socially constructed norms, since our early

---

* The arbitrariness of profanity might be important for its social purposes. Other arbitrary things like hair or clothing styles are able to signify social meaning—and to have meaning that changes over time—precisely because they're arbitrary.

childhood. Three-quarters of Americans believe in hell.[20] Most have believed in it since childhood, and we perpetuate the same belief in children in large part by reenacting the same routines that led us to believe those things in the first place: describing hell's uniquely unpleasant conditions and identifying those acts of an ill-behaved child that will land her there. We recreate hell for our children in the same way that we create word taboos for them. Early indoctrination into beliefs has effects that persist through adulthood.*

This creates a remarkable paradox. Our parents and the culture we grew up in programmed us to suppress profanity. But our reenactment of these same suppressive responses as adults gives profanity the power it has. Excessive licking exacerbates the wound. Profanity is a monster of our own perpetual creation.

$ % !

For the most part, this all seems rather harmless. The book-length love letter to profanity that you're holding in your hands might have tipped you off to the special affection I have for swearing. I think that in general we shouldn't worry so much about profanity. Aside from the special value it holds for science—we learn things about language, the mind, the brain, and society from profanity that we simply couldn't know if we pointed our microscopes elsewhere—it has practical social benefits.

For one thing, swearing increases your tolerance for pain. In several recent studies, people performed something called the "cold pressor," a task in which they're instructed to stick a hand into very cold water (five degrees centigrade) and keep it there until they can't bear it any longer.[21] This measures their pain tolerance. What happens when people swear

---

* Some people even extend these taboos into mystical realms. The Chinese word for the number four, *si*, is homonymous with the word meaning "death." As a consequence of this chance overlap in sound, many Chinese people are superstitious about saying the number four. The same is true in Japanese. It turns out that these superstitions about taboo words can have serious consequences. A paper in the *British Medical Journal* compared death figures among Chinese and Japanese Americans from 1973 to 1998 with those of Caucasians. They counted how many deaths occurred in each group on each day of the month. And they found that Chinese and Japanese Americans had significantly more heart-related deaths on the fourth day of each month than you would expect by chance, whereas the Caucasian deaths on the fourth showed no such peak. Some people are so superstitious about word taboos that they are literally scared to death. (Note though that no such increase in deaths exists for Friday the thirteenth in Western countries.)

during the cold-pressor test? In one experiment, people were told to say a swearword (like *shit*) or a control word (a word that could be used to describe a table, like *wood*) during the task. And sure enough, the people told to swear tolerated the painfully cold water significantly longer than those told to say the control word. Intriguingly, male subjects in the swearing condition also reported that the cold water was less painful than did those in the nonswearing condition, though women in the two groups reported that it was equally painful.*

And profanity has other proposed possible benefits. Timothy Jay, for instance, notes a long history of theorizing that profanity may have cathartic benefits.[22] Perhaps a well-placed *fuck* can alleviate anger that would otherwise come out in the form of physical or interpersonal aggression. Maybe swearing while you drive actually makes you less likely to take out your anger or frustration through your driving, thereby making you and those around you safer. Jay also notes that people often report feeling better after swearing[23] and that the place of profanity in some humor (so-called working blue) suggests that it might create an experience of relief. Moreover, there's evidence that some people perceive the use of profanity as a valuable social tool. One study reports that people—more so men than women—find that profanity demonstrates social power and makes the person who utters it more socially acceptable.[24]

For the most part, as we've seen, profanity does no harm. A fleeting *fuck yeah!* fades in comparison with things that demonstrably harm children and adults. There's good reason to believe that children might not be well served by exposure to violent or pornographic images until they're old enough to digest them, and of course no one could reasonably object to working to protect children from abuse, including verbal abuse. It's important for us to understand how these potential dangers affect children, their health, their development, and their relationships, and to the extent that we know they cause harm, it's worth advocating for safer environments that allow children to thrive. But to the best of our knowledge, profanity (aside from slurs, which I'll return to in a moment) leaves no such

---

* Why does swearing alleviate pain? One hypothesis is that it creates an elevated feeling of aggression. It's known that people who are more aggressive have higher tolerances for pain, and it could be that swearing hooks into the brain systems for aggression. People might be swearing themselves into a state of high pain tolerance. Swearing is special in terms of how it works in the brain, and in this case the automatic physiological reactions you have to uttering profane words allow you to better tolerate pain.

fingerprint on the child's psyche or future. Take the word *fuck*. How exactly can hearing *fuck* hurt a child? Proponents of censorship often claim that the language is "strong" or "offensive" or "immoral," but as we've seen, this means nothing other than that they themselves are offended by it or believe others might be. And there's no evidence that profanity of the Holy, Fucking, or Shit varieties harms children.

But not all profanity is equal, and all signs point to a strengthening in the United States of one specific class of profane language, namely, slurs. As Richard Dooling wrote in the *New York Times*, "We are caught between taboos. Vulgar sexual terms have become acceptable in the last two decades while all manner of sexual or ethnic epithets have become unspeakable."[25] That change is visible in the offensiveness ratings we saw earlier—where slurs are perched atop the offensiveness leaderboard. So what should we make of this drift in usage from Holy-, Fucking-, and Shit-type words to Nigger-type words?

We should care. Slurs may be the exception to the harmlessness of swearing. As we saw in the last chapter, overhearing *faggot* or *nigger* describing homosexuals or African Americans leads people to treat them as less human and to retreat from them physically. This is a type of harm. It's possible that a carefully placed *nigger* or *bitch* has the potential to lower a person's performance on tasks that his or her group is stereotyped as not good at. We saw that being called *fag* or *homo*, as an indirect, general term of offense, correlates with middle schoolers' reports of feeling less connected to their school lives and experiencing more symptoms of anxiety and depression.

Consequently, the shift away from *fuck* and toward *nigger* is a bit troubling. *Holy*, *fucking*, and *shit* were good dirty fun. But we're now giving the most power to slurs, the words that, although the evidence is still a little murky, threaten the greatest harm, at least when used in particular ways in certain contexts. So what's a socially conscious person to do? Certainly we don't want to inflict harm on anyone, especially children or people who belong to socially marginalized groups.

To add insult to injury, the rise in the prominence of slurs is a wound we're inflicting on ourselves. Our resistance to words gives them their muscle, and slurs are no exception. This means that we can't ban or censor our way out of this situation.

And other sorts of linguistic engineering seem just as quixotic as censorship. The reappropriation strategy—flooding the linguistic market with positive uses of the same word—has historically worked in certain cases; adoption and co-optation by the relevant communities may have blunted the damage that *queer* and *gay* cause. But it hasn't taken the sting out of *nigger* or *slut*. Those trying to take control over their own labels would be well served to remember that reappropriation isn't a silver bullet.

Instead—call this professor predictable—if you want to make slurs a little less powerful, you would meet with more success if everyone knew a little more about how they work and what they do. You know, better living through education.

Let's start with the supply side. Is it possible that a little knowledge would lead people to use slurs less, at least in ways known to cause harm? Here are some avenues. The first involves the low-hanging fruit. Some well-meaning people inadvertently use slurs. For example, many people use words like *gypped*, *Jewed*, or *tard* because they simply haven't ever noticed the relationship between *gypped* and *gypsy*, for example, or because they don't know the history of *tard* and *retarded*. So they have no reason to think Romani people, Jews, or people with cognitive impairments could take offense at the words. Such people don't have to be told twice why *Jewing someone* might be offensive or that *Redskins* has a history as a slur before they turn on their linguistic heels.

The second is the slightly more complicated case in which a slur gets used indirectly—still as a pejorative but not in direct reference to the original group. For example, when a twelve-year-old boy calls an opponent in an online video game *gay* or *fag*, he probably means to insult him and call him weak, but he may have no specific thoughts about his target's sexuality.[26] This is a more complicated scenario because the slur user can adopt a line of defensive reasoning that rejects any connection between a word like *fag* intended as a slur and its use in other ways. Suppose you use *fag* or *gay* or *bitch*, for example, not as slurs for homosexual people or women but rather as general terms of offense. You can easily convince yourself that you're not using sexist or homophobic language. After all, you might think to yourself, I merely called my dog a *fag*, and I know very well that he's straight from how he tries to mount the bitches at the dog park. You tell yourself, I'm not using the word to describe him as homosexual, and that's proof that I'm not using it as a gay slur. Nothing wrong with that, you might conclude.

But here's the counterpoint. We know that no matter how you use *fag*, it will bring up negative connotations. That's why slurs are often conscripted as more generic insults—using a slur for one group to insult others works only if you implicitly think the slur is derogatory in the first place. But slurs generalized beyond their original scope often still retain certain stereotypical features of their original target. Calling a man a *bitch* isn't a generic insult; it could imply that he's weak or emotional or has other attributes perceived as more stereotypically feminine. And critically embedded in this is the assumption that having these attributes is bad. So using these terms indirectly perpetuates the idea that people in the defamed group are themselves somehow bad and that having stereotypical characteristics of that group is bad. There's a lot here that could give offense.

And what's more, this may not be the most damaging use of slurs. We know that many such words crop up more often in conversations that don't include members of the targeted groups at all[27]—the majority of Caucasians use *nigger* a lot more when African Americans aren't within earshot. But this is precisely where the consequences of outsider slur use will be felt. We've seen that exposure of members of majority groups to slurs affects the way they treat and think about members of the maligned minority group: *fag* biases heterosexuals against homosexual people, for instance. So the issue with slurs isn't just what you say around the people the slur defames. It's also what you say when potentially maligned people aren't around.

I've talked through the consequences of slur use with many, many college-age people in a class I teach on profanity. Some remain undeterred and report that *fag* is unlikely to disappear from their vocabulary. But others are swayed by the evidence. And remember, a conversation with me, someone who's not popular in the slightest, accomplished this. One tweet from Kobe Bryant on the use of *gay* could change literally thousands of interactions. All things being equal, most people would prefer not to do harm, and I find that people who use slurs indirectly usually simply haven't understood the possible consequences. So that's the second way that a little knowledge might change how slurs are used.

Here's the third—reserved for the most intransigent people, those who simply like to offend. You might know these people. You might be these people. If so, you might use slurs for ideological reasons—you might see their use as a free-speech issue. And you're legally in the right. Profanity, including slurs, is protected speech. If you wanted to, you could walk down the street talking about *niggers* and *cunts* as much as you wanted without legal ramifications. Or maybe you enjoy the rise you get out of others when you slip *cunt* or *nigger* into conversation. I get it. And I get that even you, staunchest slur supporter, may possibly never be swayed by evidence. But it's also possible that the following argument for linguistic self-determination will convince you.

Here's how it goes: As Supreme Court Justice Oliver Wendell Holmes is credited with writing but probably didn't,[28] "The right to swing my fist ends where the other man's nose begins." I think this adage aptly characterizes how we think about individual rights in a lawful society. Translated to language, you have the legal right to say whatever you want, as long as it doesn't cause harm to the next person. A society clearly has a right—and a duty—to outline norms for behavior so that individuals don't hurt one another. Language is powerful, and it makes sense that some of its uses should be regulated. That's why the Supreme Court has repeatedly upheld limitations on free speech for harmful language, like libel, slander, fighting words, threats, perjury, and so on. And so, if it can be demonstrated that slurs cause harm, and there's some evidence that they do, then the fist-nose principle applies. Even if the courts haven't caught up, calling someone *nigger* or *bitch* is the linguistic analog of closing your eyes and swinging in full knowledge that there's a nose within arm's reach. The fact that you can swing doesn't mean you should.

With a little more evidence, this could become a legal argument, but I think it's still compelling as a moral one. If using language is likely to cause others harm, then maybe don't do it?

Instead, it seems respectful to refer to people and address them using the words they want to be called by. Let's call this the principle of linguistic self-determination. It's fertile for abuse, of course, and it could become time-consuming (I want to be called a *Featherless-Bipedal-Lacto-Vege-Merican!*), but most people aren't absurdists. And what's more, we know the strategy of allowing people linguistic self-determination works because it has succeeded in the replacement of *Negro* and *colored* with *black* and *African American*, of *retarded* with *developmentally challenged*, and so on. These efforts take time, but they do demonstrably create lasting change. It's possible that self-determination is a compelling enough principle to sway even some of the most hard-core slur users.

To be clear, I'm not arguing for a ban on slurs (because, as we saw in the last chapter, that's ill conceived and counterproductive), and I'm not arguing for violent knee-jerk reactions (in person or online) to people who do use slurs, even when they do so in ways that we believe may cause harm. I'm saying something quite different. In a free society, people are and should be allowed to make their own linguistic choices. But where you have a legitimate difference of opinion (about language or anything), you're also free to use the power of reason to try to persuade. That's what I'm doing right now.

And that's also where we get to the other side of the coin. It's not just that people who utter slurs might be convinced to use them a little less, at least in ways that denigrate. People who hear them might also be persuaded to temper their reaction. It's natural for people to have hair-trigger responses to words. (Anytime someone says *nigger*, it's a hate crime! Anyone who says *bitch* should be fired!) And yet, it seems like many of us would be less likely to draw immediate offense, and might enjoy better relationships with more people and more diverse ones at that, if we could overcome our immediate associations with strong words. Slurs might be offensive, but they are only words after all. Because they are neither sticks nor stones, they have to pass through the filter of our brains to cause us harm. And it's possible that knowing a little about how ephemeral and precarious their power is would allow people to worry somewhat less about superficial things like the words others are choosing. That might create more time for mindful magnanimity—assigning more import to people's actions and intentions than their word choices.

I guess this goes under the rubric of "tolerance," but not in the traditional sense. We all tolerate linguistic choices that we disagree with—even the ones we find most vile. Perhaps people who find slurs infuriating can see them for what they are: mere words. Just like any other word, each slur has a history and a future, and neither of these is the same as their present. The knee-jerk reaction to suppress these words is as unlikely a strategy as any to bear fruit. And that's probably not the worst thing in the world. After all, as George Carlin said,

> There is absolutely nothing wrong with any of those words in and of themselves. They're only words. It's the context that counts. It's the user. It's the intention behind the words that makes them good or bad. The words are completely neutral. The words are innocent.[29]

Epilogue

What Screwed the Mooch?

On July 21, 2017, Anthony Scaramucci was appointed White House communications director. Five days later, he famously placed a call to *New Yorker* reporter Ryan Lizza, and over the course of a conversation about White House leaks and other palace intrigue,[1] he let loose with a now well-documented series of colorful excursions into taboo language. He described then White House Chief of Staff Reince Priebus as "a fucking paranoid schizophrenic, a paranoiac." He contrasted himself with the then White House chief strategist, saying, "I'm not Steve Bannon. I'm not trying to suck my own cock. I'm not trying to build my own brand off the fucking strength of the President." And so on.

The published record of that conversation precipitated his departure from the White House later that week. This made for the shortest-ever term for a White House communications director—so short, in fact, that he swore his way out of a job before he was even sworn in.

And that's because of how people who swear are perceived. Some interpreted Scaramucci's profanity as indicating emotional or personality issues. The *Atlantic* described him as "too loose a cannon" who "self-destructed."[2] *Vanity Fair* called him "unhinged."[3] But impressions weren't uniformly negative. On The_Donald, a Reddit hangout for Trump fans, members opined that Scaramucci's dirty mouth showed that he "tells it like it is," that he was "unafraid."[4] Trump himself reportedly "loved the Mooch quotes" because he "likes people with backbone."[5] And the White House press secretary explained, "He's a passionate guy and sometimes he lets that passion get the better of him."[6] When he was ousted, some

Trump supporters lamented that the next communications capo would be smoother but less "real." Perceptions of swearers mix positive with negative.

Perception matters throughout our social lives. It affects how our personal relationships begin, change, and end. As Scaramucci's quick ouster shows, it also affects our likelihood of professional advancement and continued employment. Scaramucci was not the first person to get the sack for uttering profanity on the job. Alexis Hansen, a waitress at a Hooters restaurant in Ontario, California, was let go in 2013 after she lost a bikini contest and swore at the winner, who she believed had cheated.[7] In April 2013, Bismarck, North Dakota, NBC affiliate KFYR fired journalist A. J. Clemente from his new job as anchor when—perhaps unaware his microphone was on—he began his very first broadcast on his very first day of work by saying, "Fucking shit," as his first two words on live television.[8] That was also his very last day of work.

The rules—mostly unspoken—that govern workplace language are fascinating and complex. To begin with, not all workplaces are the same. I've been told by dozens of journalists that whatever the rules are about on-air language, behind the scenes, newsrooms are as vulgar as rooms can be. I've observed firsthand that construction sites can be home to unfettered profanity. I learned captivating new profanity—bilingual to boot!—by listening to the construction crew who did some recent work on our house. And yet, by contrast, it's hard to imagine much profanity slipping out at a nunnery or preschool. And even within the same workplace, the linguistic expectations when employees are in front of the public—delivering meals or the news—are far more formal and exigent than when they're in the break room. High school teachers are held to a different standard when they're standing in front of a classroom than when they're complaining among themselves in the staff room.

So perception is important, but it depends on context, and context is complicated. Swearing is a potential mine field. And in what follows, I'll try to navigate through it just a bit. We know a little about what people will think about you when you swear. Does it make them think you're more honest? More passionate? We know far less about whether those perceptions are actually true—whether swearers actually *are* more honest. But in the social world, perception may be more important than reality.

$ % !

Let's begin by tackling the issue of what swearing actually, truly says about you. If you trust science headlines from early 2017, then you might agree with fans of The Mooch, who believed his profanity was a sign of honesty. A research paper published in January of that year argued that people who swear actually tend to be more honest.[9] If true, this is a big deal for all of us, not just forensic scientists and military interrogators. But let's dig into the details.

That 2017 study asked people to estimate how often they swore and also how often they did a variety of things generally agreed to be desirable, such as keeping their promises and so on. Those who reported swearing more also reported fewer desirable traits. That's right—more swearing correlated with more *un*desirable traits. The researchers then pulled off an acrobatic piece of reasoning. They assumed that all people—swearers and nonswearers alike—have bad habits in the same measure. If that's true, they continued, then people who report more bad habits are being more honest about their negative traits. So, on this interpretation, since swearers reported more undesirable traits, they must have been responding more honestly.

There are several reasons to be skeptical of this line of reasoning. Of course, the assumption it builds in about swearers and nonswearers having negative attributes to the same degree may or may not be correct. Maybe this study shows precisely that people who swear simply have lots of other, undesirable personality traits. Maybe they don't clean up their rooms or floss their teeth. But the researchers' argument also contains a logical paradox. They conclude from this study that people who report swearing less are also less honest. But if they're less honest, then this means that we shouldn't trust their self-reports about swearing frequency as much. If we follow this thread, then people who reported low frequency of swearing were more likely to be lying about this (like they were about the other, negative attributes), so they may in fact swear just as much as the people who copped to swearing. In other words, we're back to our pet peeve from earlier chapters about self-reported, correlational studies. Maybe there's no relationship between swearing and honesty, except that people who are honest about swearing are honest about other things as well.

The way out of this loop is of course to measure honesty directly, and that's what another study did.[10] The researchers told volunteers to pick heads or tails on a coin and explained that if the coin came up on that side twice, the volunteer would receive $7. The volunteers were then instructed to flip the coin twice in complete privacy. When they came out of seclusion, they reported their coin flip results, potentially collecting their winnings if they claimed to have won, and also answered questions, including one about how strongly they agreed or disagreed with the statement "I never swear."

The researchers found, first of all, that almost everyone disagreed with this statement. Their subjects swung sweary. But there was still substantial variability in how strongly people disagreed with the statement that they never swore. Some disagreed strongly, others only somewhat. The people who were ambivalent about swearing claimed a prize about 30 percent of the time, which is just barely higher than what you'd expect if they were being honest about their choice for the two coin flips (which, based on the probability of winning two successive coin flips, is 25 percent). But the emphatic swearers claimed a prize far more often, a whopping 50 percent of the time. A rate this high suggests that some of them must have been lying to earn the extra cash. And we know about how many. Only half of the emphatic swearers claimed a prize that they actually deserved. The other half lied.

There are lots of caveats to add to this study. Swearing wasn't measured directly, so people who strongly disagreed that they didn't swear (our "emphatic swearers") could very well have differed from others in the study in a variety of ways, including their personality characteristics, how they felt in the moment, and so on. And the one thing we know about them is that they were, as a group, much more likely to lie about a prize. So it could be that they were also more likely to lie about how much they swore—perhaps as a form of braggadocio.

And even if the effects reflected real differences in lying between emphatic swearers and others, we naturally can't conclude from this experiment that swearing is *always* a sign of a dishonest character. At the very least, the relationship between swearing and honesty appears murky. Maybe it relates to honesty; maybe it doesn't. Scaramucci's swearing doesn't necessarily mean that he's any more honest than the next, more linguistically conservative person to take his place. Nor does my swearing or your swearing mean anything about how honest we are.

There's also a second prevailing belief about swearing: that it indicates lack of education or vocabulary. This charge wasn't often levied at Scaramucci, who had worked his way from a middle-class upbringing to Harvard Law. Yet it's a common complaint, especially about swearing in children—won't it stunt their vocabularies? Again, there's little evidence for this idea and a decent amount of counterevidence. For instance, one study asked participants to perform two tasks.[11] One was a measure of verbal fluency, known as the Controlled Oral Word Association Test (COWAT). This is a pretty simple task. The participant is simply told to say as many words starting with a specific letter—*a* or *s* or *f*—as he or she can in one minute, excluding proper names and never repeating the same word. On average, people came up with about fourteen words. Then they were asked to say as many swear words as they could in a minute. If swearers know fewer words in the language in general or have a harder time accessing them in real time than nonswearers, then there should be an inverse relationship between general vocabulary and profanity. That is, people who produce more swear words should produce fewer general vocabulary words in the COWAT. But the researchers found the opposite: people who produced more profane words also on average came up with more general words. This replicated across three separate studies. So the punchline is that, at least as far as we can tell, people who access more profanity more quickly are more fluent with the remainder of language. Fluency seems to be fluency, regardless of what part of the vocabulary you look at—profane or not.

A third common assumption—that by swearing, people demonstrate lower intelligence—also doesn't stand up to evidence. A recent study out of the University of Rochester—be warned, it's another correlational one— found that swearing around others is associated with "intellect." What's "intellect"? Well, these researchers had 1,065 home owners in Eugene, Oregon, fill out a standard personality questionnaire (called the "Big Five") that classified them according to certain traits, like extraversion, agreeableness, and the relevant one: intellect, which is also called "openness." Participants then rated how often they engaged in a list of four hundred different everyday acts: things like checking out a library book, eating spicy food, or swearing around other people. And the researchers found that, on average, people who rated higher on intellect said they swore more frequently.

All the normal correlational caveats apply. It's possible that smarter people don't swear more but are simply more honest about how often they swear—maybe they're better at seeing through the conventional wisdom that swearing is a sign of low intelligence. And what's more, the Big Five doesn't actually measure intelligence directly. As I mentioned, it measures "intellect," or, roughly speaking, how intellectually open or curious people say they are. Most researchers think it relates to measured intelligence, but it turns out it might not—some studies find that it doesn't correlate at all with results on intelligence tests.[12] So, in sum, there's no evidence that swearers are less intelligent than nonswearers. At the same time there's weak evidence that they might be higher on the personality trait of "intellect," whatever that ends up meaning.

The only thing that swearing has been demonstrated experimentally to say about you is that you might be in a heightened state of emotional arousal. As we saw in Chapter 9, after playing a violent, aggressive video game, people were able to name more swear words than after playing a golfing game. While people certainly swear in many situations without being emotionally aroused—to forge social intimacy, for example—all things being equal, if you compare a person who's swearing with one who's not, the one who's swearing is more likely to be emotionally aroused.

But notice what kind of information this provides about a swearer. Emotional arousal is a particular state. Anyone can be in a certain state at any given time. But to date, swearing hasn't been shown to tell us anything definitive about a person's enduring traits—personality, intelligence, honesty, or anything else. The only thing you learn for sure about a person's character by observing him or her swearing in a particular situation is that he or she is the type of person who may swear in such situations.

$ % !

And yet swearing still shifts the way people perceive you. Over the last fifty years, several dozen studies have asked subjects to make judgments about people who swear compared to those who don't. The typical setup is to have half of the subjects read a passage or watch a video in which someone expresses an opinion using profanity, while the other half watch a version with the profanity removed or replaced with neutral words. And then they rate the person for trustworthiness, politeness, pleasantness, competence, and so on.

When Anthony Scaramucci's comments came to light, reactions clustered around several responses. To generalize: people thought that he was probably expressing what he honestly felt—fervent disdain for certain other members of the administration—and that he held those beliefs intensely. Honesty and intensity. These are very common and experimentally detectable impressions of swearers.

First, honesty. In a 2005 study, researchers gave female Dutch undergrads written testimony from a burglary suspect in which he denied his involvement in the crime.[13] Half read a version that included profanity; the other half read a version with the profanity removed. And then they rated how credible they found the suspect's denial. Those who read the swearing testimony found it significantly more credible. The researchers found the same effect when they manipulated not the suspect's testimony but the victim's. When the victim swore ("That asshole pulled my bag out of my hands"), subjects were more likely to think that he was telling the truth than when he didn't ("That man pulled my bag out of my hands").

Swearers are also generally perceived as more intense. Researchers at Northern Illinois University presented eighty-eight introductory psychology students with a video in which a man argues that tuition should be lowered at another university.[14] But once again, there were two different versions of the video that either used profanity or not. Afterward, the participants were asked to rate the man's intensity and also to say whether they agreed with him. The participants who saw him swear rated him as significantly more intense and they were more likely to agree with his position. So not only can swearing make people judge you as more honest, but in some cases it can also persuade them to agree with your position.

Of course, swearing can also bring with it a host of negative perceptions. When asked directly and in the abstract what they think about those who swear, people regularly report that they find swearers untrustworthy, incompetent, and vulgar.[15] And as discussed earlier, swearers may be judged less intelligent and less articulate. Yet these judgments disappear when people's opinions are measured in reaction to a specific person swearing in context. So it seems that people merely think they judge swearers as less honest; they don't actually feel that way in the moment. Unsurprisingly—when put on the spot—we say that people who use "bad" words are bad in a variety of other ways, but we act differently when a friend, or a crime victim, or someone of seemingly passionate conviction swears. Many of our beliefs about swearing differ from how we actually behave. And that's not surprising. We've all been collecting baggage about the badness of taboo language from childhood onward.

Complicating perceptions of swearers is the finding that these perceptions differ by context. Men tend in some studies to engender stronger positive reactions when they swear publicly than women do.[16] What's more, hearing a person swear will be persuasive if you're already inclined to agree with him or her but not if you are predisposed to disagree.[17] And a stand-up comic who uses profanity extensively during a performance will surely elicit a different reaction when he swears during his day job as a kindergarten teacher. So perceptions are real but weak and fickle.

$ % !

This isn't a self-help book. But it does seem appropriate to say a few words—and just a few—about how you and I can use what we know about perceptions of swearers in our lives. Can we safeguard our personal and professional relationships while still using the peppery language that we're comfortable with? Can we exploit the perceptions that people have of swearers to produce desirable social outcomes?

I've had to confront these questions myself. I'm lucky to have an employer that values freedom of expression. Thanks to the University of California, San Diego, I can teach a class about profanity. I can write scientific papers about it. I can run experiments in my lab that expose students to it (after they consent, of course). And I was able to write a book about it. But even on the linguistically liberal West Coast and even in the heart of academia, clear strictures still limit how I can use profanity. If I swear at someone in a way that constitutes verbal abuse or sexual harassment, then, like Scaramucci, the waitress, and the anchor, I could lose my job, tenure be damned.

And yet I do choose not just to study swearing but also, in moments of my professional life, also to swear. I swear around the proverbial watercooler (in my department, it's a coffee cart) and at the end of the day over drinks with my colleagues. I swear in my lab when struggling intently to solve a problem with a graduate student. I find that using profanity in these ways makes the interactions more casual and personal and ultimately leads to stronger relationships with the people I work with. I'm probably not saying anything too controversial here. There's research showing that in informal work situations, like these, people generally find swearing unsurprising and find the swearer socially competent.[18]

In a way, I'm doing just what a researcher by the name of Stuart Jenkins did, when he took a temp job as a packer in a warehouse. In a 2007 paper,[19] he and his coauthor, Yehuda Baruch, reported that Jenkins found himself mostly excluded from social group activities for the first several months that he worked there. Then he took a chance and started working blue. From then on, they report, he was welcomed as part of the pack.

This doesn't mean, of course, that all workplace settings are suitable for alliance building through profanity. How other people are already using language in a setting usually provides the best cue to whether profanity is suitable and will be socially productive. Like other norms, every social group has expectations about the use of language. If other people of your social status are swearing in informal, non-client-facing contexts, you might be well served to join in.

Now, I also do something a little more ambitious with profanity. I actually swear in certain of my classes when I'm lecturing. Yes, I realize the pitfalls of doing so—and possibly even more so of admitting it in writing. Yes, I know that there's a perception that use of taboo language can lead to a hostile workplace environment. Yes, I know that swearing probably isn't first on the list of expectations California's taxpayers have for the education students get in my classroom. Yes, I know all about the tenured faculty who have been fired for use of profanity in lectures.

But here's what else I know. On the whole, executed with care, swearing in the classroom does little harm and some measurable good. That's not just my impression. Here's the evidence.

To begin with, remember that I'm teaching adults. College students. If I drop a fleeting expletive, I'm not saying anything they haven't already seen a hundred times that morning before class on social media. And empirical work has shown that students find instructor swearing inoffensive to a degree that most people would never imagine. A 2009 study asked sixty-seven college students at a small, private Midwestern university to rate fifty-six potentially offensive classroom behaviors an instructor might exhibit—things like cutting off a student who is talking, not making eye contact with students, or swearing.[20] Even I was startled that swearing was rated fifty-first out of the fifty-six behaviors—less offensive than everything from hitting on a student, of course (third most offensive), to talking in a monotone (forty-first most offensive). The only items on the list rated less offensive than swearing were talking about his or her personal life, offering a strong opinion, grading on a curve, ending class early, and drinking a beverage while teaching. Eating while teaching was more offensive than swearing. So was reading PowerPoint slides. In short, swearing was not so offensive.

That doesn't mean that all swearing in the classroom is judged equally. College students judge certain uses of swearing in the classroom, and different targets of that swearing, as more appropriate than others. Researchers asked 272 students at a large southwestern public university to recall an instance when an instructor swore in class, then to describe it and rate how appropriate it was.[21] Students found certain uses of profanity inappropriate—for instance, to show frustration or to be more "relatable." But they found it quite appropriate for the instructor to use swearing for emphasis, to gain attention, and to be humorous. Moreover, as long as the swearing was directed toward course content (like a particularly challenging idea) or tools of the trade (like a computer not working during the instructor's presentation), they found the use of taboo language on average somewhat appropriate. But when it was directed at students or at a particular assignment, the appropriateness ratings dropped. Other work comes to the same conclusion about profanity. When it's used as a form of verbal aggression toward students, it counterproductively leads students to find instructors less caring and compassionate and to feel more negatively about not only the instructor but the course content as well.[22] Swearing is globally positive but only when targeted appropriately.

What's the result for me, personally? Students say in their student evaluations that they overwhelmingly find me accessible, honest, and funny, and upward of 90 percent of the time, they would recommend me to other students. Now, I don't let any of this go to my head because I know they're just judging me on the basis of how I appear to them and 299 of their classmates from behind a lectern on a stage. In truth, the actual me may or not be accessible, honest, or funny. (In the interest of transparency, the answers in order are not really, about average, and you be the judge.) And yet I don't believe I'm cynically using profanity as a kind of tactical deception. Yes, I'm exploiting the predictability of perceptions about swearers. But I'm also just being myself. I like to make fun of my fumbling failures with PowerPoint using a well-placed *dumbfuck*. It feels right to me to focus the class's attention on one particularly crucial point of an argument using an unexpected *good shit*. It doesn't just make them feel better about me. It makes me feel more comfortable with them.

Similar studies haven't been conducted for every occupation. Professors study the language of professors professing more than they study that of masons or CPAs or designers. So for other professions, you may need to do your own fieldwork. Get the lay of the land. Observe the reactions to taboo language when others use it. Start low on the chart (*shit* is a good entry-level swear) before moving up the ladder. And as with anything in the workplace, err on the side of conservatism.

As a result of being a person who wrote a book about swearing, people often ask me what it means when someone swears. As you see, in general terms, we don't really know. So for the time being, in my humble opinion, we might be better served by steering clear of evaluating people because they swear or because they don't. Not just Anthony Scaramucci. Also me. People's words might not mean anything about their personalities or their experiences. For that matter, people might be tactically deceiving us with their swears, like moths who wear colorful patches on their wings that look like gigantic eyes to frighten off potential predators.[23] We'd do better to not judge people by the colorfulness of their words, which says little about the content of their character.

Acknowledgments

I began writing this book as a way to procrastinate while working on another project. Like other stalling techniques, this book soon came to occupy more of my time than what I was actually supposed to be doing. You are now holding in your hands the product of seven years of committed procrastination. There's some satisfaction in leveraging one's own character flaws against themselves.

Although I was the chief procrastinator, this book is not a product of my making alone. Writing a book is hard in many of the ways that writing other things is hard. Words don't lend themselves easily to clarity, especially when the thoughts behind them are muddled. And I have many people to thank for helping me think and write more clearly. At the outset, I owe a debt of gratitude to my writing group collaborators: Loriena Yancura, Katherine Irwin, and Ashley Maynard. Even if it's been said before, it bears repeating that laboring over each word is essential; it yields better results when you do it with other people, and you end up a better writer for it too. If we hadn't laughed so much reading early chapter drafts, I might not have stuck with the book.

I've also been fortunate to have smart, energetic students—most of them now colleagues—read the book in part or in its entirety and help shape the argument, the ideas, and the writing. Thanks to Tyler Marghetis, Kensy Cooperrider, Ross Metusalem, Josh Davis, and Arturs Semenuks. I also discussed the book along the way with—and got ideas that you'll find in these pages from—Nancy Chang, Hans Boas, Lera Boroditsky, Scott Klemmer, David Kirsh, Jawee Perla, Iris Kohlberg, Madelaine Plauché, Elizabeth Moyer, and Adam Ruderman.

Writing a book is also hard in its own particular ways. Books cover ground beyond one's area of confident expertise, and a number of experts in the various fields the book's chapters touch on were kind enough to share insights into their areas of research or provide feedback on chapters. A hearty thanks to Timothy Jay, Eric Bakovic, Roger Levy, Seana Coulson, Marta Kutas, Michael Motley, Donna Jo Napoli, Michael Israel, Elinor Ochs, Piotr Winkielman, Karen Dobkins, Andrea Carnaghi, Taylor Jones, Christopher Hall, Al Schutz, Alison Gopnik, Vic Ferreira, Matthew Fisher, and Diana Van Lancker Sidtis. These people didn't merely deliver platitudes and tell me the work was fine. They told me what I got wrong, and the book is better for it. One person even held the considered and expert opinion that I got everything in one chapter wrong and said in so many words that the book wasn't worth the cost of its ink. I told this esteemed colleague that I'd express my gratitude for his candor by refraining from sending him a copy of the book, and I intend to go through with that promise, but I would be remiss not to also thank Roger Lass on this page.

Books are also especially hard because they compete in a marketplace of unlike kinds—with not just other trade science or nonfiction books, or even books in general, but also with basically all gift and entertainment products. So they have to be appealing, and they have to be accessible. I can't imagine anyone better equipped to take the bird's-eye view and lay out the path to writing a book that will have broad appeal than Katinka Matson. She and the staff at Brockman helped craft the immature product of a procrastinating mind into something worth showing to anyone. And in the same vein, T. J. Kelleher at Basic Books has been my Charon through the production process. How fortunate I've been to have an editor with such a very similar sense of humor and aesthetics to mine. And if I do say so, T. J. has impeccable taste.

Books are also hard to write because, concretely, doing so requires many thousands of hours. In my case, those hours were virtually all spent on the couch in my living room, many of them in long stretches. This makes the living room largely unlivable for other members of the household, not merely because of the droning electronica "for focusing" but also due to the inevitable squalor. I want to thank Frances for pretending that this is a normal state of affairs and for accommodating my appropriation of the living room. She has also helped me work through specific ideas, chapter organization, and writing details. She didn't write this book, but I wouldn't have been able to write it without her. Thank you to poop-machine Matthew and to the rest of my family—David, Karen, Dori, Ira, Joshua, Seila, Dave, Caroline, Jon, and everyone else—for your support and love throughout.

Notes

Introduction
1. Lussier, C. (June 27, 2015).
2. E.g., Jay, T. (1999).
3. Pinker, S. (2007).

Chapter 1: *Holy, Fucking, Shit, Nigger*
1. Stephens, R., Atkins, J., and Kingston, A. (2009).
2. Napoli, D. J., and Hoeksema, J. (2009).
3. Jay, T. (1999).
4. Hughes, G. (1998).
5. Federal Communications Commission (n.d.).
6. Broadcasting Standards Authority (2010).
7. Millwood-Hargrave, A. (2000).
8. Thanks, Katherine Tse and Alexsandra McMahan!
9. Janschewitz, K. (2008). See also its conceptual predecessor: Jay, T. (1992).
10. Pang, C. (2007).
11. Nechepurenko, I. (May 5, 2015).
12. Voutilainen, E. (2008).
13. Hughes, G. (2006).
14. McGrath, P., and Phillips, E. (2008).
15. Frazer, J. (1922).
16. Mohr, M. (2013).
17. Bonnefous, B. (September 15, 2014).
18. Lefton, B. (August 29, 2014).
19. Bergen, B. K. (2012).

Chapter 2: What Makes a Four-Letter Word?
1. Lewis, M. P., Simons, G. F., and Fennig, C. D. (2009).
2. Gabelentz, G. v. d. (1891).

3. Thibodeau, P. H., Bromberg, C., Hernandez, R., and Wilson, Z. (2014).
4. OED Online, "Glost."
5. Bergen, B. K. (2004).
6. Velten, H. D. V. (1943).

Chapter 3: One Finger Is Worth a Thousand Words

1. Nida, E. A. (1949).
2. Rebouças, C. B. D. A., Pagliuca, L. M. F., and Almeida, P. C. D. (2007).
3. Alaska Shorthand Reporters Association (2015).
4. Malcolm, A. (April 17, 2008).
5. Songbass (November 3, 2008).
6. Wafflesouls (August 8, 2012).
7. Lebra, T. S. (1976).
8. Cooperrider, K., and Núñez, R. (2012).
9. Chandler, J., and Schwarz, N. (2009).
10. Center for Information Dominance: Center for Language, Regional Expertise, and Culture (2010).
11. Snopes (October 11, 2014).
12. Aristophanes (1968) and Nasaw, D. (February 6, 2012).
13. Robbins, I. P. (2008).
14. Laertius, D. (1925).
15. Marsh, P., Morris, D., and Collett, P. (1980).
16. Robbins, I. P. (2008).
17. Farmer, J. S. (1890).
18. Pennington, J. (July 27, 2014).
19. Axtell, R. (1998).
20. Nasaw, D. (February 6, 2012).
21. Nasaw, D. (February 6, 2012).
22. Matsumoto, D., and Hwang, H. S. (2012).
23. Link, M. (July 26, 2010).
24. Marsh, P., Morris, D., and Collett, P. (1980).
25. Brentari, D. (2011).
26. Liddell, S. K. (1980).
27. Mirus, G., Fisher, J., and Napoli, D. J. (2012).
28. Mirus, G., Fisher, J., and Napoli, D. J. (2012), 1005.
29. Mirus, G., Fisher, J., and Napoli, D. J. (2012), 1007.
30. Mirus, G., Fisher, J., and Napoli, D. J. (2012), 1011.
31. Taub, S. F. (2001).
32. Frishberg, N. (1975).
33. Mitchell, R. E., Young, T. A., Bachleda, B., and Karchmer, M. A. (2006).
34. Lewis, M. P., Simons, G. F., and Fennig, C. D. (2009).
35. Baker-Shenk, C. L., Baker, C., and Padden, C. (1979).

36. Bellugi, U., and Fischer, S. (1972); Bellugi, U., Fischer, S., and Newkirk, D. (1979).

37. Napoli, D. J., Fisher, J., and Mirus, G. (2013).

38. Deuchar, M. (2013).

39. Lieberth, A. K., and Gamble, M. E. B. (1991); Baus, C., Carreiras, M., and Emmorey, K. (2013).

40. Xu, J., Gannon, P. J., Emmorey, K., Smith, J. F., and Braun, A. R. (2009).

Chapter 4: The Holy Priest with the Vulgar Tongue

1. Bogousslavsky, J., Hennerici, M. G., Bäzner, H., and Bassetti, C. (Eds.) (2010).

2. Lecours, A. R., Nespoulous, J. L., and Pioger, D. (1987).

3. Lordat, J. (1843).

4. Translation by Lecours, A. R., Nespoulous, J. L., and Pioger, D. (1987).

5. Liechty, J. A., and Heinzekehr, J. B. (2007).

6. Here's an exception: Jay, T. (1999).

7. Finger, S. (2001).

8. Raichle, M. E., and Gusnard, D. A. (2002).

9. Friederici, A. D. (2011).

10. Gazzaniga, M. S. (2004).

11. Van Lancker, D., and Cummings, J. L. (1999).

12. Lancker, D. V., and Nicklay, C. K. (1992).

13. Jay, T. (1999).

14. Kosslyn, S. M., and Miller, G. W. (2013).

15. Knecht, S., Drager, B., Deppe, M., Bobe, L., Lohmann, H., Floel, A., Ringelstein, E.-B., and Henningsen, H. (2000).

16. Graves, R., and Landis, T. (1990).

17. Code, C. (1996).

18. Smith, A. (1966).

19. Speedie, L. J., Wertman, E., Ta'ir, J., and Heilman, K. M. (1993).

20. Chakravarthy, V. S., Joseph, D., and Bapi, R. S. (2010).

21. Peterson, B., Riddle, M. A., Cohen, D. J., Katz, L. D., Smith, J. C., Hardin, M. T., and Leckman, J. F. (1993); Singer, H. S., Reiss, A. L., Brown, J. E., Aylward, E. H., Shih, B., Chee, E., Harris, E. L., Reader, M. J., Chase, G. A., Bryan, R. N., and Denckla, M. B. (1993).

22. Jankovic, J., and Rohaidy, H. (1987).

23. Friedman, S. (1980).

24. Van Lancker, D., and Cummings, J. L. (1999).

25. Robinson, B. W. (1972); Myers, R. E. (1976); Pinker, S. (2007); Jay, T. (1999).

26. Robinson, B. W. (1967); Jurgens, U., and Ploog, D. (1970).

27. Jay, T. (1999); Pinker, S. (2007).

28. Jackson, J. H. (1974/1958), cited by Jay, T. (1999).

Chapter 5: The Day the Pope Dropped the *C*-Bomb

1. Dicker, R. (March 3, 2014).
2. Chappell, B. (March 3, 2014).
3. Pisa, N. (March 3, 2014).
4. Erard, M. (2008).
5. Miller, N., Maruyama, G., Beaber, R. J., and Valone, K. (1976).
6. Fromkin, V. (1980).
7. Nooteboom, S. G. (1995).
8. Chappell, B. (March 3, 2014).
9. Fromkin, V. A. (Ed.) (1984).
10. Freud, S. (1966 [1901]).
11. *Huffington Post* Staff (May 10, 2010).
12. Schoeneman, D. (April 26, 2004).
13. Motley, M. T., and Baars, B. J. (1979).
14. Pincott, J. (March 13, 2012).
15. Mahl, G. F. (1987); Baars, B. J. (Ed.) (1992).
16. Poulisse, N. (1999).
17. Motley, M. T., Camden, C. T., and Baars, B. J. (1981).
18. McGinnies, E. (1949).
19. Horvath, F. (1978).
20. Harris, C. L., Aycicegi, A., and Gleason, J. B. (2003).
21. Motley, M. T., Camden, C. T., and Baars, B. J. (1981).
22. Severens, E., Janssens, I., Kühn, S., Brass, M., and Hartsuiker, R. J. (2011).
23. Kutas, M., and Federmeier, K. D. (2011).
24. Severens, E., Janssens, I., Kühn, S., Brass, M., and Hartsuiker, R. J. (2011).
25. Severens, E., Kühn, S., Hartsuiker, R. J., and Brass, M. (2012).
26. Aron, A. R., Robbins, T. W., and Poldrack, R. A. (2014).
27. Siegrist, M. (1995).
28. LaBar, K. S., and Phelps, E. A. (1998).
29. Mackay, D. G., Shafto, M., Taylor, J. K., Marian, D. E., Abrams, L., and Dyer, J. R. (2004).
30. MacKay, D. G., and Ahmetzanov, M. V. (2005).
31. Schriefers, H., Meyer, A. S., and Levelt, W. J. (1990).
32. Dhooge, E., and Hartsuiker, R. J. (2011).

Chapter 6: *Fucking* Grammar

1. Carlin, G. (1990).
2. Shad, U. P. (1992).
3. Postal, P. (2004).
4. Horn, L. R. (1997).
5. Quang, P. D. (1971 [1992]).
6. Hoeksema, J., and Napoli, D. J. (2008).

7. Hoeksema, J., and Napoli, D. J. (2008).
8. Daintrey, L. (1885), cited in Hoeksema, J., and Napoli, D. J. (2008).
9. Goldberg, A. E. (1995).

Chapter 7: How *Cock* Lost Its Feathers

1. Liuzza, R. M. (1994).
2. OED Online, "Cock."
3. OED Online, "Dick."
4. Social Security Administration (n.d.).
5. Lass, R. (1995).
6. Lass, R. (1995).
7. Medievalists.net (September 10, 2015).
8. OED Online, "Clock."
9. Partridge, E., and Beale, P. (1984).
10. OED Online, "Bitch."
11. Altmann, E. G., Pierrehumbert, J. B., and Motter, A. E. (2011).
12. Eisenstein, J., O'Connor, B., Smith, N. A., and Xing, E. P. (2014).
13. Munro, P. (Ed.) (1993).
14. Cheadle, W. (2010).
15. Figure from Eisenstein, J., O'Connor, B., Smith, N. A., and Xing, E. P. (2014).
16. Lass, R. (1995).
17. Holt, R., and Baker, N. (2001).
18. von Fleischhacker, R. (1894 [1400]).
19. Coşkun, R. (December 5, 2013).
20. Klepousniotou, E. (2002).
21. Williams, J. N. (1992).
22. Winslow, A. G. (1900 [1772]).
23. OED Online, "Motherfucker."
24. Franklyn, J. (Ed.) (2013).
25. Oxford Dictionaries (June 22, 2011).

Chapter 8: Little Samoan Potty Mouths

1. *Telegraph* Staff (January 4, 2010).
2. Ochs, E. (1982).
3. Aubrey, A. (March 27, 2008).
4. Aubrey, A. (March 27, 2008).
5. Davis, B. L., and MacNeilage, P. F. (1995).
6. Kern, S., and Davis, B. L. (2009).
7. Lenneberg, E. H., Rebelsky, F. G., and Nichols, I. A. (1965).
8. Speech and Language Pathology Department of the Center for Communication at the Children's Hospital of Philadelphia (June 2, 2011).

9. Lee, S. A. S., Davis, B., and MacNeilage, P. (2010).
10. Lenneberg, E. H., Rebelsky, F. G., and Nichols, I. A. (1965).
11. Oller, D. K., and Eilers, R. E. (April 1988).
12. Petitto, L. A., and Marentette, P. F. (1991).
13. Lee, S. A. S., Davis, B., and MacNeilage, P. (2010).
14. Davis, B. L., and MacNeilage, P. F. (1995).
15. Hunkin, G. A. (2009).
16. Trask, L. (2004).
17. Murdock, G. P. (1959).
18. Jakobson, R. (1962).
19. Ochs, E. (1982).
20. Ochs, E. (1982), 19.

Chapter 9: Fragile Little Minds

1. Bologna, C. (August 15, 2014).
2. City of North Augusta, South Carolina (October, 2010).
3. *Huffington Post* (n.d.).
4. Rivenburg, R. (February 28, 2007).
5. Srisavasdi, R. (March 1, 2007).
6. American Academy of Pediatrics (October 17, 2011).
7. Gilani, N. (October 17, 2011).
8. Shepherd, R. (October 18, 2011).
9. Welsh, J. (October 17, 2011).
10. Coyne, S. M., Stockdale, L. A., Nelson, D. A., and Fraser, A. (2011).
11. Coyne, S. M., Stockdale, L. A., Nelson, D. A., and Fraser, A. (2011).
12. Griffiths, M. D., and Shuckford, G. L. J. (1989).
13. Coyne, S. M., Stockdale, L. A., Nelson, D. A., and Fraser, A. (2011).
14. Buchanan, T. W., Etzel, J. A., Adolphs, R., and Tranel, D. (2006).
15. Shea, S. A. (1996).
16. Lin, I. M., and Peper, E. (2009).
17. Peltonen, K., Ellonen, N., Larsen, H. B., and Helweg-Larsen, K. (2010).
18. Choi, S., and Gopnik, A. (1995).
19. American National Corpus (n.d.).
20. Goodman, J. C., Dale, P. S., and Li, P. (2008).
21. Roy, B. C., Frank, M. C., and Roy, D. (2009).
22. Schwartz, R., and Terrell, B. (1983).
23. Jay, K. L., and Jay, T. B. (2013).
24. Karen Dobkins reminded me of this demonstration.
25. Zile, A., and Stephens, R. (2014).
26. Nisbett, R. E., and Wilson, T. D. (1977).
27. Thorndike, E. L. (1920).
28. Verhulst, B., Lodge, M., and Lavine, H. (2010).
29. Leuthesser, L., Kohli, C. S., and Harich, K. R. (1995).

Chapter 10: The $100,000 Word
1. Bresnahan, M. (April 13, 2011).
2. Malec, B., and Nessif, B. (May 2, 2013).
3. Maske, M. (July 29, 2014).
4. Loy, J. W., and Elvogue, J. F. (1970).
5. Kain, D. J. (2008).
6. Byers, W., and Hammer, C. (1997).
7. Janschewitz, K. (2008).
8. Hughes, G. (1998).
9. Johnson, C. (October 14, 1904).
10. Yung, J., Chang, G. H., and Lai, H. M. (Eds.) (2006).
11. Fritz Pollard Alliance (April 13, 2014).
12. Fasoli, F., Maass, A., and Carnaghi, A. (2014).
13. Fasoli, F., Paladina, M. P., Carnaghi, A., Jetten, J., Bastian, B., and Bain, P. (in press).
14. Fasoli, F., Paladina, M. P., Carnaghi, A., Jetten, J., Bastian, B., and Bain, P. (in press).
15. Carnaghi, A., and Maass, A. (2007).
16. Carnaghi, A., and Maass, A. (2007).
17. Poteat, V. P., and Espelage, D. L. (2007).
18. Steele, C. M. (2010).
19. Shih, M., Ambady, N., Richeson, J. A., Fujita, K., and Gray, H. M. (2002).
20. Fritz Pollard Alliance (November 21, 2013).
21. National Association for the Advancement of Colored People (July 9, 2007).
22. 2PacDonKilluminati (April 23, 2010).
23. Galinsky, A. D., Wang, C. S., Whitson, J. A., Anicich, E. M., Hugenberg, K., and Bodenhausen, G. V. (2013).
24. Jones, T. W., and Hall, C. S. (March 2015).
25. Spears, A. K. (1998).
26. Spears, A. K. (1998).
27. Goodwin, M. H. (2002).
28. Highkin, S. (November 14, 2013).
29. Chalabi, M. (April 28, 2014).
30. Smith, M. (March 3, 2014).

Chapter 11: The Paradox of Profanity
1. O'Connor, J. (2006).
2. Jay, K., and Jay, T. (2015).
3. Jay, T. B., King, K., and Duncan, T. (2006).
4. Nalkur, P. G., Jamieson, P. E., and Romer, D. (2010).
5. Gentile, D. A. (2008).
6. Ivory, J. D., Williams, D., Martins, N., and Consalvo, M. (2009).

7. Federal Communications Commission (n.d.).
8. *FCC v. Pacifica Foundation*, 438 U.S. 726 (1978).
9. Jay, T. (2009a).
10. Kaye, B. K., and Sapolsky, B. S. (2004).
11. Wachal, R. S. (2002).
12. Nye, J., Ferreira, F., Husband, E. M., and Lyon, J. M. (2012).
13. Kremar, M., and Sohn, S. (2004).
14. *Guardian* (July 2, 1999).
15. TVTropes (n.d.).
16. Zeitzen, M. (2008).
17. Albrecht, H.-J., and Sieber, U. (2007).
18. Joint Monitoring Programme (2014).
19. Holm, B. R., Rest, J. R., and Seewald, W. (2004).
20. Gallup (n.d.).
21. Stephens, R., Atkins, J., and Kingston, A. (2009).
22. Jay, T. (2009b).
23. Jay, T. B., King, K., and Duncan, D. (2006).
24. Selnow, G. W. (1985).
25. Dooling, R. (September 6, 1996).
26. Warner, J. (April 16, 2009).
27. Graumann, C. F. (1995).
28. Quote Investigator (October 15, 2011).
29. Carlin, G. (1990).

Epilogue: What Screwed the Mooch?

1. Lizza, R. (July 27, 2017).
2. Graham, D. A. (July 31, 2017).
3. Levin, B. (July 27, 2017).
4. I vastly prefer profanity-laced TRUTH to sweet, fluffy lies, thread posted to Reddit by sergeimagnitsky. Retrieved from www.reddit.com/r/The_Donald /comments/6q3ie8/i_vastly_prefer_profanitylaced_truth_to_sweet.
5. Allen, M. (July 28, 2017).
6. Baker, P., and Haberman, M. (July 27, 2017).
7. Korkki, P. (February 13, 2016).
8. Anchor, A. J. (April 23, 2013).
9. Feldman, G., Lian, H., Kosinski, M., and Stillwell, D. (January 15, 2017).
10. de Vries, R. E., Hilbig, B. E., Zettler, I., Dunlop, P. D., Holtrop, D., Lee, K., and Ashton, M. C. (July 21, 2017).
11. Jay, K. L., and Jay, T. B. (2015).
12. Furnham, A., Moutafi, J., and Chamorro-Premuzic, T. (2005).
13. Rassin, E., and Van Der Heijden, S. (2005).
14. Scherer, C. R., and Sagarin, B. J. (2006).

15. Bostrom, R. N., Baseheart, J. R., and Rossiter, C. M. (1973); Hamilton, M. A. (1989); Paradise, L. V., Cohl, B., and Zweig, J. (1980); Cavazza, N., and Guidetti, M. (2014).
16. Cavazza, N., and Guidetti, M. (2014).
17. Bostrom, R. N., Baseheart, J. R., & Rossiter, C. M. (1973).
18. Johnson, D. I., and Lewis, N. (2010).
19. Baruch, Y., and Jenkins, S. (2007).
20. Stork, E., and Hartley, N. T. (2009).
21. Generous, M. A., Frei, S. S., and Houser, M. L. (2015).
22. Myers, S. A., and Knox, R. L. (1999); Myers, S. A. (2001).
23. Byrne, R., and Whiten, A. (1991).

Works Cited

2PacDonKilluminati. (April 23, 2010). 2Pac interview with Tabitha Soren Part 1. YouTube. Retrieved from https://www.youtube.com/watch?v =ljSCZyv97FY.

Alaska Shorthand Reporters Association. (2015). *Making the record*. Retrieved from http://www.alaskashorthandreporters.org/index.php?option =com_content&view=article&id=6&Itemid=4.

Albrecht, H.-J., and Sieber, U. (2007). Stellungnahme zu dem Fragenkatalog des Bundesverfassungsgerichts in dem Verfahren 2 BvR 392/07 zu § 173 Abs. 2 S. 2 StGB—Beischlaf zwischen Geschwistern. Max Planck Institute for Foreign and International Criminal Law, p. 29.

Allen, M. (July 28, 2017). Trump "loved" Scaramucci's quotes—but he hates being upstaged. *Axios*. Retrieved from www.axios.com/trump-loved -scaramuccis-quotes-but-he-hates-being-upstaged-2466546181.html.

Altmann, E. G., Pierrehumbert, J. B., and Motter, A. E. (2011). Niche as a determinant of word fate in online groups. *PLOS ONE*, 6(5), e19009.

American Academy of Pediatrics. (October 17, 2011). Profanity in the media linked to youth swearing, aggression. Retrieved from http://www.aap .org/en-us/about-the-aap/aap-press-room/Pages/Profanity-in-the-Media -Linked-to-Youth-Swearing,-Aggression.aspx.

American National Corpus. (n.d.). ANC second release frequency data. Retrieved from http://www.anc.org/SecondRelease/data/ANC-all-count.txt.

Anchor, A. J. (April 23, 2013). Clemente fired after profanity-laced debut. ABC News. Retrieved from http://abcnews.go.com/blogs/entertainment/2013 /04/anchor-a-j-clemente-fired-after-profanity-laced-debut.

Apatow, J. (Producer), and McKay, A. (Director). (2004). *Anchorman: The legend of Ron Burgundy*. Universal City, CA: DreamWorks SKG.

Aristophanes. (1968). *The clouds*. K. J. Dover (Ed.). Oxford: Oxford University Press.

Aron, A. R., Robbins, T. W., and Poldrack, R. A. (2014). Inhibition and the right inferior frontal cortex: One decade on. *Trends in Cognitive Sciences*, 18(4), 177–185.

Aubrey, A. (March 27, 2008). Why kids curse. National Public Radio. Retrieved from http://www.npr.org/templates/story/story.php?storyId=89127830.

Axtell, R. (1998). *Gestures: The do's and taboos of body language around the world*. New York: John Wiley & Sons, Inc.

Baars, B. J. (Ed.). (1992). *Experimental slips and human error: exploring the architecture of volition*. New York: Plenum Press.

Baillet, S. (2014). Forward and inverse problems of MEG/EEG. In *Encyclopedia of computational neuroscience*, D. Jaeger and R. Jung (Eds.), 1–8. New York: Springer.

Baker, P., and Haberman, M. (July 27, 2017). Anthony Scaramucci's uncensored rant: Foul words and threats to have Priebus fired. *New York Times*. Retrieved from www.nytimes.com/2017/07/27/us/politics/scaramucci-priebus-leaks.html?mcubz=1.

Baker-Shenk, C. L., Baker, C., and Padden, C. (1979). *American Sign Language: A look at its history, structure, and Community*. Silver Spring, MD: TJ Publishers.

Baruch, Y., and Jenkins, S. (2007). Swearing at work and permissive leadership culture: When anti-social becomes social and incivility is acceptable. *Leadership & Organization Development Journal*, 28(6), 492–507.

Baus, C., Carreiras, M., and Emmorey, K. (2013). When does iconicity in sign language matter? *Language and Cognitive Processes*, 28(3), 261–271.

Bellugi, U., and Fischer, S. (1972). A comparison of sign language and spoken language. *Cognition*, 1, 173–200.

Bellugi, U., Fischer, S., and Newkirk, D. (1979). The rate of speaking and signing. In *The signs of language*, E. Klima and U. Bellugi (Eds.), 181–194. Cambridge, MA: Harvard University Press.

Bergen, B. K. (2004). The psychological reality of phonaesthemes. *Language*, 80(2), 290–311.

Bergen, B. K. (2012). *Louder than words: The new science of how the mind makes meaning*. New York: Basic Books.

Bogousslavsky, J., Hennerici, M. G., Bäzner, H., and Bassetti, C. (Eds.). (2010). *Neurological disorders in famous artists*. Basel: Karger Medical and Scientific Publishers.

Bologna, C. (August 15, 2014). South Carolina mom arrested for cursing in front of her kids. *Huffington Post*. Retrieved from http://www.huffingtonpost.com/2014/08/15/mom-arrested-for-swearing_n_5681837.html.

Bonnefous, B. (September 15, 2014). Manuel Valls droit dans ses bottes face à sa majorité. *Le Monde*. Retrieved from http://www.lemonde.fr/politique/article/2014/09/15/manuel-valls-droit-dans-ses-bottes-a-sa-majorite_4487435_823448.html.

Bostrom, R. N., Baseheart, J. R., and Rossiter, C. M. (1973). The effects of three types of profane language in persuasive messages. *Journal of Communication*, *23*, 461–475.

Brentari, D. (2011). Sign language phonology. In *The handbook of phonological theory*, J. A. Goldsmith, J. Riggle, and A. C. L. Yu (Eds.), 691–721. 2nd ed. Malden, MA: Wiley-Blackwell.

Bresnahan, M. (April 13, 2011). Lakers' Kobe Bryant is fined $100,000 by NBA for anti-gay slur to referee. *Los Angeles Times*. Retrieved from http://articles .latimes.com/2011/apr/13/sports/la-sp-kobe-bryant-lakers-20110414.

Broadcasting Standards Authority. (2010). *What not to swear: The acceptability of words in broadcasting*. Retrieved from http://bsa.govt.nz/images/assets /Research/What-Not-to-Swear-Full-BSA2010.pdf.

Buchanan, T. W., Etzel, J. A., Adolphs, R., and Tranel, D. (2006). The influence of autonomic arousal and semantic relatedness on memory for emotional words. *International Journal of Psychophysiology*, *61*(1), 26–33.

Byers, W., and Hammer, C. (1997). *Unsportsmanlike conduct: Exploiting college athletes*. Ann Arbor: University of Michigan Press.

Byrne, R., and Whiten, A. (1991). Computation and mindreading in primate tactical deception. In *Natural theories of mind: Evolution, development and simulation of everyday mindreading*. Whiten, A. (Ed.), 127–141. Cambridge, UK: Basil Blackwell.

Carlin, G. (1990). *Parental advisory—explicit lyrics*. New York: Eardrum Records.

Carnaghi, A., and Maass, A. (2007). In-group and out-group perspectives in the use of derogatory group labels: Gay versus fag. *Journal of Language and Social Psychology*, *26*(2), 142–156.

Cavazza, N., and Guidetti, M. (2014). Swearing in political discourse: Why vulgarity works. *Journal of Language and Social Psychology*, *33*(5), 537–547.

Center for Information Dominance: Center for Language, Regional Expertise, and Culture. (2010). Behavior and etiquette—other physical gestures: Beckoning and American gestures. Islamic Republic of Afghanistan. University of West Florida. Retrieved from http://uwf.edu/atcdev/Afghanistan/Behaviors /Lesson8PhysicalGestures.html.

Chakravarthy, V. S., Joseph, D., and Bapi, R. S. (2010). What do the basal ganglia do? A modeling perspective. *Biological Cybernetics*, *103*(3), 237–253.

Chalabi, M. (April 28, 2014). Three Leagues, 92 teams and one black principal owner. *FiveThirtyEight*. Retrieved from http://fivethirtyeight.com/datalab /diversity-in-the-nba-the-nfl-and-mlb.

Chandler, J., and Schwarz, N. (2009). How extending your middle finger affects your perception of others: Learned movements influence concept accessibility. *Journal of Experimental Social Psychology*, *45*(1), 123–128.

Chappell, B. (March 3, 2014). Pope Francis lets a vulgarity slip during Vatican address. National Public Radio. Retrieved from http://www.npr.org

/blogs/thetwo-way/2014/03/03/285271511/pope-francis-lets-a-vulgarity-slip
-during-vatican-address.

Cheadle, W. (2010). *Cheadle's journal of trip across Canada, 1862–1863*. Victoria, BC: TouchWood Editions.

Choi, S., and Gopnik, A. (1995). Early acquisition of verbs in Korean: A cross-linguistic study. *Journal of Child Language, 22*(3), 497–529.

Chomsky, N. (1965). *Aspects of the theory of syntax*. Cambridge, MA: MIT Press.

City of North Augusta, South Carolina. (October, 2010). Code of Ordinances. Supplement No. 25. Retrieved from https://law.resource.org/pub/us/code/city /sc/North%20Augusta,%20SC%20Code%20thru%20Supp%20%2325.pdf.

Code, C. (1996). Speech from the isolated right hemisphere? Left hemispherectomy cases EC and NF. In *Classic cases in neuropsychology*, C. Code, Y. Joanette, A. R. Lecours, and C. W. Wallesch (Eds.), 291–307. Hove, UK: Psychology Press.

Conlee, J. (2004). *William Dunbar: The complete works*. Consortium for the Teaching of the Middle Ages (TEAMS). Kalamazoo, MI: Medieval Institute Publications.

Cooperrider, K., and Núñez, R. (2012). Nose-pointing. *Gesture, 12*(2).

Coşkun, R. (December 5, 2013). Nice pussy TV anchor blooper. YouTube. Retrieved from https://www.youtube.com/watch?v=QjgW631T5Tc.

Coyne, S. M., Stockdale, L. A., Nelson, D. A., and Fraser, A. (2011). Profanity in media associated with attitudes and behavior regarding profanity use and aggression. *Pediatrics, 128*(5), 867–872.

Daintrey, Laura. (1885). *The king of Alberia: A romance of the Balkans*. London: Methuen and Company.

Davis, B. L., and MacNeilage, P. F. (1995). The articulatory basis of babbling. *Journal of Speech, Language, and Hearing Research, 38*(6), 1199–1211.

Deuchar, M. (2013). *British Sign Language*. London: Routledge.

de Vries, R. E., Hilbig, B. E., Zettler, I., Dunlop, P. D., Holtrop, D., Lee, K., and Ashton. M. C. (July 21, 2017). Honest people tend to use less—not more— profanity: Comment on Feldman et al.'s (2017) study 1. *Social Psychological and Personality Science*. Retrieved from http://journals.sagepub.com/doi /full/10.1177/1948550617714586.

Dhooge, E., and Hartsuiker, R. J. (2011). How do speakers resist distraction? Evidence from a taboo picture-word interference task. *Psychological Science, 22*(7), 855–859.

Dicker, R. (March 3, 2014). Oops! Pope Francis accidentally says the F-word in Italian. *Huffington Post*. Retrieved from http://www.huffingtonpost .com/2014/03/03/pope-cazzo-francis-says-f-word_n_4890328.html.

Dooling, R. (September 6, 1996). Unspeakable names. *New York Times*. Retrieved from http://www.nytimes.com/1996/09/06/opinion/unspeakable -names.html.

Dowd, M. (January 31, 1999). Liberties; niggardly city. *New York Times*, 17.

Eisenstein, J., O'Connor, B., Smith, N. A., and Xing, E. P. (2014). Diffusion of lexical change in social media. *PLOS ONE*, 9(11), e113114.

Erard, M. (2008). Um . . . : Slips, stumbles, and verbal blunders, and what they mean. *Anchor Canada*, 243.

Farmer, J. S. (1890). *Slang and its analogues past and present: A dictionary, historical and comparative of the heterodox speech of all classes of society for more than three hundred years* (Vol. 1). London: Printed for subscribers only.

Fasoli, F., Maass, A., and Carnaghi, A. (2014). Labelling and discrimination: Do homophobic epithets undermine fair distribution of resources? *British Journal of Social Psychology* 54(2), 383–393.

Fasoli, F., Paladina, M. P., Carnaghi, A., Jetten, J., Bastian, B., and Bain, P. (In press). Not "just words": Exposure to homophobic epithets leads to dehumanizing and physical distancing from gay men. *European Journal of Social Psychology*.

FCC v. Pacifica Foundation, 438 U.S. 726 (1978).

Federal Communications Commission. (n.d.). Obscene, indecent and profane broadcasts. Retrieved from https://consumercomplaints.fcc.gov/hc/en-us/articles/202731600-Obscene-Indecent-and-Profane-Broadcasts.

Feldman, G., Lian, H., Kosinski, M., and Stillwell, D. (January 15, 2017). Frankly, we do give a damn: The relationship between profanity and honesty. *Social Psychological and Personality Science*. Retrieved from http://journals.sagepub.com/doi/full/10.1177/1948550616681055.

Fillmore, C. J. (1985). Syntactic intrusions and the notion of grammatical construction. In *The Eleventh Annual Meeting of the Berkeley Linguistics Society*, M. Niepokuj, M. VanClay, V. Nikiforidou, and D. Feder (Eds.), 73–86. Berkeley, CA: Berkeley Linguistics Society.

Finger, S. (2001). *Origins of neuroscience: A history of explorations into brain function*. Oxford: Oxford University Press.

Franklyn, J. (Ed.). (2013). *A dictionary of rhyming slang*. London: Routledge.

Frazer, J. (1922). Names of the dead tabooed. In *The Golden Bough*, abr. ed. New York: Macmillan Company.

Freud, S. (1966 [1901]). *The psychopathology of everyday life* (No. 611). New York: W. W. Norton and Company.

Friederici, A. D. (2011). The brain basis of language processing: From structure to function. *Physiological reviews*, 91(4), 1357–1392.

Friedman, S. (1980). Self-control in the treatment of Gilles de la Tourette's syndrome: Case study with 18-month follow-up. *Journal of Consulting and Clinical Psychology*, 48(3), 400.

Frishberg, N. (1975). Arbitrariness and iconicity: Historical change in American Sign Language. *Language*, 51(3), 696–719.

Fritz Pollard Alliance. (November 21, 2013). Time to put an end to the "N" word on NFL playing fields. Retrieved from http://fritzpollard.org/time-to-put-an-end-to-the-n-word-on-nfl-playing-fields.

Fritz Pollard Alliance. (April 13, 2014). FPA commends NFL on commitment to rid the league of offensive language. Retrieved from http://fritzpollard.org/fpa-commends-nfl-on-commitment-to-rid-the-league-of-offensive-language.

Fromkin, V. (1980). *Errors in linguistic performance: Slips of the tongue, ear, pen, and hand.* New York: Academic Press.

Fromkin, V. A. (Ed.). (1984). *Speech errors as linguistic evidence* (Janua Linguarum. Series Maior 77). Berlin: Walter de Gruyter.

Furnham, A., Moutafi, J., and Chamorro-Premuzic, T. (2005). Personality and intelligence: Gender, the Big Five, self-estimated and psychometric intelligence. *International Journal of Selection and Assessment, 13*(1), 11–24.

Gabelentz, G. v. d. (1891). Die Sprachwissenschaft, ihre Aufgaben, Methoden und bisherigen Ergebnisse. Leibzig: Chr. Herm. Tauchnitz.

Galinsky, A. D., Wang, C. S., Whitson, J. A., Anicich, E. M., Hugenberg, K., and Bodenhausen, G. V. (2013). The reappropriation of stigmatizing labels the reciprocal relationship between power and self-labeling. *Psychological Science,* 0956797613482943.

Gallup. (n.d.). Religion. Retrieved from http://www.gallup.com/poll/1690/religion.aspx.

Gazzaniga, M. S. (2004). *The cognitive neurosciences.* Cambridge, MA: MIT Press.

Generous, M. A., Frei, S. S., and Houser, M. L. (2015). When an instructor swears in class: Functions and targets of instructor swearing from college students' retrospective accounts. *Communication Reports, 28*(2), 128–140.

Gentile, D. A. (2008). The rating systems for media products. In *The handbook of children, media, and development,* Sandra L. Calvert and Barbara J. Wilson (Eds.), 527–551. Malden, MA: Wiley-Blackwell, 2008.

Gilani, N. (October 17, 2011). Bad language in video games and on TV linked to aggression in teenagers. *Daily Mail.* Retrieved from http://www.daily mail.co.uk/news/article-2050159/Bad-language-video-games-TV-linked-aggression-teenagers.html.

Goldberg, A. E. (1995). *Constructions: A construction grammar approach to argument structure.* Chicago: University of Chicago Press.

Goodman, J. C., Dale, P. S., and Li, P. (2008). Does frequency count? Parental input and the acquisition of vocabulary. *Journal of Child Language, 35*(3), 515.

Goodwin, M. H. (2002). Building power asymmetries in girls' interaction. *Discourse & Society, 13*(6), 715–730.

Graham, D. A. (July 31, 2017). The spectacular self-destruction of Anthony Scaramucci. *Atlantic.* Retrieved from www.theatlantic.com/politics/archive/2017/07/seven-against-thebes/535464.

Graumann, C. F. (1995). Discriminatory discourse. *Pattern of Prejudice, 29,* 69–83.

Graves, R., and Landis, T. (1985). Hemispheric control of speech expression in aphasia: A mouth asymmetry study. *Archives of Neurology, 42*(3), 249–251.

Graves, R., and Landis, T. (1990). Asymmetry in mouth opening during different speech tasks. *International Journal of Psychology, 25*(2), 179–189.

Griffiths, M. D., and Shuckford, G. L. J. (1989). Desensitization to television violence: A new model. *New Ideas in Psychology, 70*(1), 85–89.

Guardian. (July 2, 1999). Movies. Retrieved from http://www.theguardian.com/film/1999/jul/02/news.

Hamilton, M. A. (1989). Reactions to obscene language. *Communication Research Reports, 6*, 67–69.

Harris, C. L., Aycicegi, A., and Gleason, J. B. (2003). Taboo words and reprimands elicit greater autonomic reactivity in a first language than in a second language. *Applied Psycholinguistics, 24*(4), 561–579.

Highkin, S. (November 14, 2013). Charles Barkley says Matt Barnes should not apologize for racial slur. *USA Today*. Retrieved from http://ftw.usatoday.com/2013/11/nba-charles-barkley-matt-barnes-racial-slur.

Hoeksema, J., and Napoli, D. J. (2008). Just for the hell of it: A comparison of two taboo-term constructions. *Journal of Linguistics, 44*(2), 347–378.

Holm, B. R., Rest, J. R., and Seewald, W. (2004). A prospective study of the clinical findings, treatment and histopathology of 44 cases of pyotraumatic dermatitis. *Veterinary Dermatology, 15*(6), 369–376.

Holt, R., and Baker, N. (2001). Indecent exposure—sexuality, society and the archaeological record. In *Towards a geography of sexual encounter: Prostitution in English medieval towns*, L. Bevan (Ed.). Glasgow: Cruithne Press.

Horn, L. R. (1997). Flaubert triggers, squatitive negation, and other quirks of grammar. *Tabu, 26*, 183–205.

Horvath, F. (1978). An experimental comparison of the psychological stress evaluator and the galvanic skin response in detection of deception. *Journal of Applied Psychology, 63*(3), 338.

Huffington Post. (n.d.) Weird South Carolina. *Huffington Post*. Retrieved from http://www.huffingtonpost.com/news/weird-south-carolina.

Huffington Post Staff. (May 10, 2010). Tiger Woods 'bulging dick': Golf Channel makes unfortunate gaffe. *Huffington Post*. Retrieved from http://www.huffingtonpost.com/2010/05/10/tiger-woods-bulging-dick_n_569792.html.

Hughes, G. (1998). *Swearing: A social history of foul language, oaths and profanity in English*. London: Penguin UK.

Hughes, G. (2006). *An encyclopedia of swearing: The social history of oaths, profanity, foul language, and ethnic slurs in the English-speaking world*. Armonk, NJ: M. E. Sharpe.

Hunkin, G. A. (2009). *Gagana Sāmoa: A Samoan language coursebook*. Honolulu: University of Hawaii Press.

I vastly prefer profanity-laced TRUTH to sweet, fluffy lies, thread posted to Reddit by sergeimagnitsky. Retrieved from www.reddit.com/r/The_Donald /comments/6q3ie8/i_vastly_prefer_profanitylaced_truth_to_sweet.

Ide, S., and Ueno, K. (2011). Honorifics and address terms. *Pragmatics of Society, 5,* 439.

Ivory, J. D., Williams, D., Martins, N., and Consalvo, M. (2009). Good clean fun? A content analysis of profanity in video games and its prevalence across game systems and ratings. *CyberPsychology and Behavior, 12*(4), 457–460.

Jackson, J. H. (1974 [1958]). *Selected writings of John Hughlings Jackson* (Vol. 1). London: Staples Press.

Jakobson, R. (1962). Why "mama" and "papa." *Selected writings,* Vol. 1: *Phonological studies,* 538–545. The Hague: Mouton.

Jankovic, J., and Rohaidy, H. (1987). Motor, behavioral and pharmacologic findings in Tourette's syndrome. *Canadian Journal of Neurological Sciences/Journal canadien des sciences neurologiques, 14*(3 Suppl), 541–546.

Janschewitz, K. (2008). Taboo, emotionally valenced, and emotionally neutral word norms. *Behavior Research Methods, 40*(4), 1065–1074.

Jay, K. L., and Jay, T. B. (2013). A child's garden of curses: A gender, historical, and age-related evaluation of the taboo lexicon. *American Journal of Psychology, 126*(4), 459–475.

Jay, K. L., and Jay, T. B. (2015). Taboo word fluency and knowledge of slurs and general pejoratives: Deconstructing the poverty-of-vocabulary myth. *Language Sciences, 52,* 251–259.

Jay, T. (1992). *Cursing in America: A psycholinguistic study of dirty language in the courts, in the movies, in the schoolyards, and on the streets.* Philadelphia: John Benjamins Publishing Company.

Jay, T. (1999). *Why we curse: A neuro-psycho-social theory of speech.* Philadelphia: John Benjamins Publishing Company.

Jay, T. (2009a). Do offensive words harm people? *Psychology, Public Policy, and Law, 15*(2), 81.

Jay, T. (2009b). The utility and ubiquity of taboo words. *Perspectives on Psychological Science, 4*(2), 153–161.

Jay, T. B., King, K., and Duncan, D. (2006). Memories of punishment for cursing. *Sex Roles, 32,* 123–133.

Johnson, C. (October 14, 1904). "They are only 'niggers' in the South." *Seattle Republican.*

Johnson, D. I., and Lewis, N. (2010). Perceptions of swearing in the work setting: An expectancy violations theory perspective. *Communication Reports, 23*(2), 106–118.

Joint Monitoring Programme. (2014). Progress on drinking water and sanitation, 2014 update. WHO/UNICEF Joint Monitoring Programme for Water Supply and Sanitation, p. 6.

Jones, T. W., and Hall, C. S. (March 2015). Semantic bleaching and the emergence of new pronouns in AAVE. In *LSA Annual Meeting Extended Abstracts, 6*.

Jurgens, U., and Ploog, D. (1970). Cerebral representation of vocalization in the squirrel monkey. *Experimental Brain Research, 10*, 532–554.

Kain, D. J. (2008). It's just a concussion: The National Football League's denial of a casual link between multiple concussions and later-life cognitive decline. *Rutgers Law Journal, 40*, 697.

Kaye, B. K., and Sapolsky, B. S. (2004). Watch your mouth! An analysis of profanity uttered by children on prime-time television. *Mass Communication and Society, 7*(4), 429–452.

Kern, S., and Davis, B. L. (2009). Emergent complexity in early vocal acquisition: Cross-linguistic comparisons of canonical babbling. In *Approaches to phonological complexity*, F. Pellegrino, E. Marsico, I. Chitoran, and C. Coupé (Eds.), 353–375. New York: Mouton de Gruyter.

Klepousniotou, E. (2002). The processing of lexical ambiguity: Homonymy and polysemy in the mental lexicon. *Brain and Language, 81*(1), 205–223.

Knecht, S., Drager, B., Deppe, M., Bobe, L., Lohmann, H., Floel, A., Ringelstein, E.-B., and Henningsen, H. (2000). Handedness and hemispheric language dominance in healthy humans. *Brain, 123*, 2512–2518.

Korkki, P. (February 13, 2016). Fired for cursing on the job, testing the limits of labor law. *New York Times*. Retrieved from www.nytimes.com/2016/02/14/business/fired-for-cursing-on-the-job-testing-the-limits-of-labor-law.html?mcubz=1.

Kosslyn, S. M., and Miller, G. W. (2013). *Top brain, bottom brain: Surprising insights into how you think*. New York: Simon and Schuster.

Kremar, M., and Sohn, S. (2004). The role of bleeps and warnings in viewers' perceptions of on-air cursing. *Journal of Broadcasting & Electronic Media, 48*(4), 570–583.

Kutas, M., and Federmeier, K. D. (2011). Thirty years and counting: Finding meaning in the N400 component of the event-related brain potential (ERP). *Annual Review of Psychology, 62*, 621.

LaBar, K. S., and Phelps, E. A. (1998). Arousal-mediated memory consolidation: Role of the medial temporal lobe in humans. *Psychological Science, 9*(6), 490–493.

Laertius, D. (1925). *Lives of eminent philosophers*, trans. R. D. Hicks. Cambridge, MA: Loeb Classical Library.

Lancker, D. V., and Nicklay, C. K. (1992). Comprehension of personally relevant (PERL) versus novel language in two globally aphasic patients. *Aphasiology, 6*(1), 37–61.

Lass, R. (1995). Four letters in search of an etymology. *Diachronica 12*, 99–111.

Lebra, T. S. (1976). *Japanese patterns of behaviour*. Honolulu: University of Hawaii Press.

Lecours, A. R., Nespoulous, J. L., and Pioger, D. (1987). Jacques Lordat or the birth of cognitive neuropsychology. In *Motor and sensory processes of language*, E. Keller and M. Gopnik (Eds.), 1–16. Hove, UK: Psychology Press.

Lee, S. A. S., Davis, B., and MacNeilage, P. (2010). Universal production patterns and ambient language influences in babbling: A cross-linguistic study of Korean- and English-learning infants. *Journal of Child Language*, 37(2), 293–318.

Lefton, B. (August 29, 2014). Ichiro Suzuki uncensored, en Español. *Wall Street Journal*. Retrieved from http://online.wsj.com/articles/ichiro -suzuki-uncensored-en-espanol-1409356461.

Lenneberg, E. H., Rebelsky, F. G., and Nichols, I. A. (1965). The vocalizations of infants born to deaf and to hearing parents. *Human Development*, 8(1), 23–37.

Leuthesser, L., Kohli, C. S., and Harich, K. R. (1995). Brand equity: The halo effect measure. *European Journal of Marketing*, 29(4), 57–66.

Levin, B. (July 27, 2017). "I want to f*cking kill all the leakers": Scaramucci explodes in unhinged late-night rant. *Vanity Fair*. Retrieved from www .vanityfair.com/news/2017/07/anthony-scaramucci-new-yorker.

Lewis, M. P., Simons, G. F., and Fennig, C. D. (2009). *Ethnologue: Languages of the world*. Dallas, TX: SIL International.

Liddell, S. K. (1980). *American Sign Language syntax* (Approaches to Semiotics 52). New York: Mouton de Gruyter.

Lieberth, A. K., and Gamble, M. E. B. (1991). The role of iconicity in sign language learning by hearing adults. *Journal of Communication Disorders*, 24(2), 89–99.

Liechty, J. A., and Heinzekehr, J. B. (2007). Caring for those without words: A perspective on aphasia. *Journal of Neuroscience Nursing*, 39(5), 316–318.

Lin, I. M., and Peper, E. (2009). Psychophysiological patterns during cell phone text messaging: A preliminary study. *Applied Psychophysiology and Biofeedback*, 34(1), 53–57.

Link, M. (July 26, 2010). Dangerous body language abroad. *AOL Travel*. Retrieved from http://news.travel.aol.com/2010/07/26/dangerous-body -language-abroad.

Liuzza, R. M. (1994). *The Old English version of the gospels*. New York: Oxford University Press.

Lizza, R. (July 27, 2017). Anthony Scaramucci called me to unload about White House leakers, Reince Priebus, and Steve Bannon. *New Yorker*. Retrieved from www.newyorker.com/news/ryan-lizza/anthony-scaramucci-called-me -to-unload-about-white-house-leakers-reince-priebus-and-steve-bannon.

Lordat, J. (1843). *Analyse de la parole, pour servir à la théorie de divers cas d'alalie et de paralalie (de mutisme et d'imperfection du parler) que les nosologistes ont mal connus*. Montpellier, FR: L. Castel.

Loy, J. W., and Elvogue, J. F. (1970). Racial segregation in American sport. *International Review for the Sociology of Sport, 5*(1), 5–24.

Lussier, C. (June 27, 2015). LSU professor fired for using salty language in classroom claims she's 'witch hunt' victim, plans suit. *Advocate*. Retrieved from http://theadvocate.com/news/12669113-123/lsu-professor-fired-for-using.

MacKay, D. G., and Ahmetzanov, M. V. (2005). Emotion, memory, and attention in the taboo Stroop paradigm: An experimental analogue of flashbulb memories. *Psychological Science, 16*(1), 25–32.

MacKay, D. G., Shafto, M., Taylor, J. K., Marian, D. E., Abrams, L., and Dyer, J. R. (2004). Relations between emotion, memory, and attention: Evidence from taboo Stroop, lexical decision, and immediate memory tasks. *Memory and Cognition, 32*(3), 474–488.

Mahl, G. F. (1987). *Explorations in nonverbal and vocal behavior*. Hillsdale, NJ: Erlbaum.

Mair, V. (September 4, 2014). The paucity of curse words in Japanese. Language Log. Retrieved from http://languagelog.ldc.upenn.edu/nll/?p=14412.

Malcolm, A. (April 17, 2008). Barack Obama makes a one-fingered gesture while speaking of Hillary Clinton. *Los Angeles Times*. Retrieved from http://latimesblogs.latimes.com/washington/2008/04/obamaflipsoffcl.html.

Malec, B., and Nessif, B. (May 2, 2013). Kobe Bryant on Jason Collins coming out as gay: "As his peers we have to support him." *EOnline*. Retrieved from http://www.eonline.com/news/414400/kobe-bryant-on-jason-collins -coming-out-as-gay-as-his-peers-we-have-to-support-him.

Marsh, P., Morris, D., and Collett, P. (1980). *Gestures, their origins and distribution*. Lanham, MD: Madison Books.

Maske, M. (July 29, 2014). NFL will have 'zero tolerance' for on-field use of racial and homophobic slurs, players are told in league's officiating video. *Washington Post*. Retrieved from https://www.washingtonpost.com/news /sports/wp/2014/07/29/nfl-will-have-zero-tolerance-for-on-field-use-of -racial-and-homophobic-slurs-players-are-told-in-leagues-officiating -video.

Matsumoto, D., and Hwang, H. S. (2012). Body and gestures. *Nonverbal Communication: Science and Applications, 75*.

Mayo, M. (November 11, 2011). Is using this N-word (niggardly) a firing offense? *Sun Sentinel*.

McGinnies, E. (1949). Emotionality of perceptual defense. *Psychological Review, 56*, 244–251.

McGrath, P., and Phillips, E. (2008). Australian findings on Aboriginal cultural practices associated with clothing, hair, possessions and use of name of deceased persons. *International Journal of Nursing Practice, 14*(1), 57–66.

Medievalists.net. (September 10, 2015). The earliest use of the F-word discovered. Retrieved from http://www.medievalists.net/2015/09/10/the-earliest -use-of-the-f-word-discovered.

Miller, N., Maruyama, G., Beaber, R. J., and Valone, K. (1976). Speed of speech and persuasion. *Journal of Personality and Social Psychology, 34*(4), 615.

Millwood-Hargrave, A. (2000). *Delete expletives?* Ofcom. Retrieved from http://www.ofcom.org.uk/static/archive/itc/uploads/Delete_Expletives.pdf.

Mirus, G., Fisher, J., and Napoli, D. J. (2012). Taboo expressions in American Sign Language. *Lingua, 122*(9), 1004–1020.

Mitchell, R. E., Young, T. A., Bachleda, B., and Karchmer, M. A. (2006). How many people use ASL in the United States? Why estimates need updating. *Sign Language Studies, 6*(3), 306–335.

Mohr, M. (2013). *Holy sh*t: A brief history of swearing.* Oxford: Oxford University Press.

Motley, M. T., and Baars, B. J. (1979). Effects of cognitive set upon laboratory induced verbal (Freudian) slips. *Journal of Speech, Language, and Hearing Research, 22*(3), 421–432.

Motley, M. T., Camden, C. T., and Baars, B. J. (1981). Toward verifying the assumptions of laboratory-induced slips of the tongue: The output-error and editing issues. *Human Communication Research, 8*(1), 3–15.

Munro, P. (Ed.). (1993). UCLA slang 2. *UCLA Occasional Papers in Linguistics, 12*, 58.

Murdock, G. P. (1959). Cross-language parallels in parental kin terms. *Anthropological Linguistics, 1*(9), 1–5.

Myers, S. A. (2001). Perceived instructor credibility and verbal aggressiveness in the college classroom. *Communication Research Reports, 18*(4), 354–364.

Myers, S. A., and Knox, R. L. (1999). Verbal aggression in the college classroom: Perceived instructor use and student affective learning. *Communication Quarterly, 47*(1), 33–45.

Nalkur, P. G., Jamieson, P. E., and Romer, D. (2010). The effectiveness of the Motion Picture Association of America's rating system in screening explicit violence and sex in top-ranked movies from 1950 to 2006. *Journal of Adolescent Health, 47*(5), 440–447.

Napoli, D. J., Fisher, J., and Mirus, G. (2013). Bleached taboo-term predicates in American Sign Language. *Lingua, 123*, 148–167.

Napoli, D. J., and Hoeksema, J. (2009). The grammatical versatility of taboo terms. *Studies in Language, 33*(3), 612–643.

Nasaw, D. (February 6, 2012). When did the middle finger become offensive? *BBC News Magazine.* Retrieved from http://www.bbc.com/news/magazine-16916263.

National Association for the Advancement of Colored People. (July 9, 2007). The "N" word is laid to rest by the NAACP. Retrieved from http://www.naacp.org/press/entry/the--n--word-is-laid-to-rest-by-the-naacp.

Nechepurenko, I. (May 5, 2015). Putin bans the F-word from movies and plays. *Moscow Times.* Retrieved from http://www.themoscowtimes.com/news/article/putin-bans-the-f-word-from-movies-and-plays/499530.html.

Nida, E. A. (1949). *Morphology: The descriptive analysis of words.* Ann Arbor: University of Michigan Press.

Nisbett, R. E., and Wilson, T. D. (1977). The halo effect: Evidence for unconscious alteration of judgments. *Journal of Personality and Social Psychology,* 35(4), 250.

Nooteboom, S. G. (1995). Limited lookahead in speech production. In *Producing speech: Contemporary issues: For Katherine Safford Harris,* K. S. Harris, F. Bell-Berti, and L. J. Raphael (Eds.), 1–18. New York: AIP Press.

Nye, J., Ferreira, F., Husband, E. M., and Lyon, J. M. (2012). Reconstruction of censored taboo in sentence processing. Poster given at the *25th CUNY Human Sentence Processing Conference,* New York, NY, March 14–16.

O'Connor, J. (2006). *Cuss control: The complete book on how to curb your cursing.* iUniverse.

Ochs, E. (1982). Talking to children in western Samoa. *Language in Society,* 11(1), 77–104.

OED Online. (September 18, 2015). Bitch. *OED Online.* http://www.oed.com/view/Entry/19524.

OED Online. (September 18, 2015). Cock. *OED Online.* http://www.oed.com/view/Entry/35327.

OED Online. (September 18, 2015). Dick. *OED Online.* http://www.oed.com/view/Entry/52255.

OED Online. (September 18, 2015). Glost. *OED Online.* http://www.oed.com/view/Entry/79187.

OED Online. (September 18, 2015). Motherfucker. *OED Online.* http://www.oed.com/view/Entry/242538.

Oller, D. K., and Eilers, R. E. (April 1988). The role of audition in infant babbling. *Child Development,* 59(2), 441–449.

Oxford Dictionaries. (June 22, 2011). The OEC: Facts about the language. Retrieved from http://oxforddictionaries.com/words/the-oec-facts-about-the-language.

Pang, C. (2007). *Little dogs are too lazy to polish shoes (小狗懶擦鞋): A study of Hong Kong profanity culture* [in Chinese]. Hong Kong: Subculture Publishing.

Paradise, L. V., Cohl, B., and Zweig, J. (1980). Effects of profane language and physical attractiveness on perceptions of counselor behavior. *Journal of Counseling Psychology,* 27, 620–624.

Partridge, E., and Beale, P. (1984). *A dictionary of slang and unconventional English.* London: Routledge.

Peltonen, K., Ellonen, N., Larsen, H. B., and Helweg-Larsen, K. (2010). Parental violence and adolescent mental health. *European Child and Adolescent Psychiatry,* 19(11), 813–822.

Pennington, J. (July 27, 2014). What is the origin of the phrase "flipping the bird"? *Quora.* Retrieved from http://www.quora.com/What-is-the-origin-of-the-phrase-flipping-the-bird.

Peterson, B., Riddle, M. A., Cohen, D. J., Katz, L. D., Smith, J. C., Hardin, M. T., and Leckman, J. F. (1993). Reduced basal ganglia volumes in Tourette's syndrome using three-dimensional reconstruction techniques from magnetic resonance images. *Neurology, 43*(5), 941–949.

Petitto, L. A., and Marentette, P. F. (1991). Babbling in the manual mode: Evidence for the ontogeny of language. *Science, 251*(5000), 1493–1496.

Phillips, D. P., Liu, G. C., Kwok, K., Jarvinen, J. R., Zhang, W., and Abramson, I. S. (2001). The Hound of the Baskervilles effect: Natural experiment on the influence of psychological stress on timing of death. *BMJ, 323*(7327), 1443–1446.

Pincott, J. (March 13, 2012). Slips of the tongue. *Psychology Today*. Retrieved from https://www.psychologytoday.com/articles/201203/slips-the-tongue.

Pinker, S. (2007). *The stuff of thought: Language as a window into human nature*. New York: Penguin.

Pisa, N. (March 3, 2014). Pope drops the F-bomb: Pontiff gets his Italian mixed up during Sunday blessing. *Daily Mail*. Retrieved from http://www.daily mail.co.uk/news/article-2572086/Pope-accidentally-says-f-Sunday-blessing -getting-Italian-wrong.html.

Postal, P. (2004). The structure of one type of American English vulgar minimizer. In *Skeptical Linguistic Essays*, 159–172. New York: Oxford University Press.

Poteat, V. P., and Espelage, D. L. (2007). Predicting psychosocial consequences of homophobic victimization in middle school students. *Journal of Early Adolescence, 27*(2), 175–191.

Poulisse, N. (1999). *Slips of the tongue: Speech errors in first and second language production* (Studies in Bilingualism 20). Philadelphia: John Benjamins Publishing Company.

Quang, P. D. (1992 [1971]). English sentences without overt grammatical subject. In *Studies out in left field: Defamatory essays presented to James D. McCawley on the occasion of his 33rd or 34th birthday*, James D. McCawley and Arnold M. Zwicky (Eds.), 3–10. Philadelphia: John Benjamins Publishing Company.

Quote Investigator. (October 15, 2011). Your liberty to swing your fist ends just where my nose begins. Retrieved from http://quoteinvestigator.com /2011/10/15/liberty-fist-nose.

Myers, R. E. (1976). Comparative neurology of vocalization and speech: Proof of a dichotomy. *Annals of the New York Academy of Science, 280*, 745–757.

Raichle, M. E., and Gusnard, D. A. (2002). Appraising the brain's energy budget. *Proceedings of the National Academy of Sciences, 99*(16), 10237–10239.

Rassin, E., and Van Der Heijden, S. (2005). Appearing credible? Swearing helps! *Psychology, Crime & Law, 11*(2). Retrieved from http://dx.doi.org/10 .1080/10683160516051233132295 2.

Rebouças, C. B. D. A., Pagliuca, L. M. F., and Almeida, P. C. D. (2007). Non-verbal communication: Aspects observed during nursing consultations with blind patients. *Escola Anna Nery*, *11*(1), 38–43.

Rivenburg, R. (February 28, 2007). She swears the Constitution is on her side. *Los Angeles Times*. Retrieved from http://articles.latimes.com/2007/feb/28/local/me-swear28.

Robbins, I. P. (2008). Digitus impudicus: The middle finger and the law. *UC Davis Law Review*, *41*, 1403–1485.

Robinson, B. W. (1967). Vocalization evoked from forebrain in Macaca mulatta. *Physiology and Behavior*, *2*(4), 345–354.

Robinson, B. W. (1972). Anatomical and physiological contrasts between human and other primate vocalizations. In *Perspectives on Human Evolution*, S. L. Washburn and P. Dolhinow (Eds.), 2, 438–443. New York: Holt, Rinehart and Winston.

Roy, B. C., Frank, M. C., and Roy, D. (2009). Exploring word learning in a high-density longitudinal corpus. In *Proceedings of the 31st Annual Meeting of the Cognitive Science Society 2009 (CogSci 2009): Amsterdam, Netherlands, 29 July–1 August 2009*, Niels A Taatgen and Henri van Rijn (Eds.). Austin, TX: Cognitive Science Society.

Sandritter, M. (October 15, 2014). Colin Kaepernick didn't use a racial slur, still fined by the NFL. *SBNation*. Retrieved from http://www.sbnation.com/nfl/2014/10/15/6985459/colin-kaepernick-fined-racial-slur-nfl-nflpa-49ers.

Scanlon, T. J., Luben, R. N., Scanlon, F. L., and Singleton, N. (1993). Is Friday the 13th bad for your health? *BMJ*, *307*(6919), 1584–1586.

Scherer, C. R., and Sagarin, B. J. (2006). Indecent influence: The positive effects of obscenity on persuasion. *Social Influence*, *1*(2), 138–146. doi: 10.1080/15534510600747597.

Schoeneman, D. (April 26, 2004). Armani's exchange . . . Condi's slip . . . forget the Alamo. *New York Magazine*. Retrieved from http://nymag.com/nymetro/news/people/columns/intelligencer/n_10245.

Schriefers, H., Meyer, A. S., and Levelt, W. J. (1990). Exploring the time course of lexical access in language production: Picture-word interference studies. *Journal of Memory and Language*, *29*(1), 86–102.

Schwartz, R., and Terrell, B. (1983). The role of input frequency in lexical acquisition. *Journal of Child Language*, *10*, 57–64.

Selnow, G. W. (1985). Sex differences in uses and perceptions of profanity. *Sex Roles*, *12*(3–4), 303–312.

Severens, E., Janssens, I., Kühn, S., Brass, M., and Hartsuiker, R. J. (2011). When the brain tames the tongue: Covert editing of inappropriate language. *Psychophysiology*, *48*(9), 1252–1257.

Severens, E., Kühn, S., Hartsuiker, R. J., and Brass, M. (2012). Functional mechanisms involved in the internal inhibition of taboo words. *Social Cognitive and Affective Neuroscience*, *7*(4), 431–435.

Shad, U. P. (1992). Some unnatural habits. In *Studies out in left field: Defamatory essays presented to James D. McCawley on the occasion of his 33rd or 34th birthday*, James D. McCawley and Arnold M. Zwicky (Eds.), 33–36. Philadelphia: John Benjamins Publishing Company.

Shea, S. A. (1996). Behavioural and arousal-related influences on breathing in humans. *Experimental Physiology, 81*(1), 1–26.

Shepherd, R. (October 18, 2011). Teen aggression increased by profanity in TV and video games. *Medical News Today*. Retrieved from http://www.medical newstoday.com/articles/236182.php.

Shih, M., Ambady, N., Richeson, J. A., Fujita, K., and Gray, H. M. (2002). Stereotype performance boosts: The impact of self-relevance and the manner of stereotype activation. *Journal of Personality and Social Psychology, 83*(3), 638.

Siegrist, M. (1995). Effects of taboo words on color-naming performance on a Stroop test. *Perceptual and Motor Skills, 81*(3f), 1119–1122.

Singer, H. S., Reiss, A. L., Brown, J. E., Aylward, E. H., Shih, B., Chee, E., Harris, E. L., Reader, M. J., Chase, G. A., Bryan, R. N., and Denckla, M. B. (1993). Volumetric MRI changes in basal ganglia of children with Tourette's syndrome. *Neurology, 43*(5), 950.

Smith, A. (1966). Speech and other functions after left (dominant) hemispherectomy. *Journal of Neurology, Neurosurgery, and Psychiatry, 29*(5), 468.

Smith, M. (March 3, 2014). Richard Sherman calls NFL banning the N-word "an atrocious idea." *NBC Sports*. Retrieved from http://profootballtalk .nbcsportscom/2014/03/03/richard-sherman-calls-nfl-banning-the-n -word-an-atrocious-idea.

Snopes. (October 11, 2014). Pluck Yew. Retrieved from http://www.snopes.com /language/apocryph/pluckyew.asp.

Social Security Administration. (n.d.). Background information. Retrieved from https://www.ssa.gov/oact/babynames/background.html.

Songbass. (November 3, 2008). Obama gives McCain the middle finger. YouTube. Retrieved from https://www.youtube.com/watch?v=Pc8Wc1CN7sY.

Spears, A. K. (1998). African-American language use: Ideology and so-called obscenity. In *African-American English: Structure, history, and use*, Salikoko S. Mufwene (Ed.), 226–250. New York: Routledge.

Speech and Language Pathology Department of the Center for Communication at the Children's Hospital of Philadelphia. (June 2, 2011). Speech sound milestones for children. SpeechandLanguage.com. Retrieved from http:// www.speechandlanguage.com/clinical-cafe/speech-sound-milestones -for-children.

Speedie, L. J., Wertman, E., Ta'ir, J., and Heilman, K. M. (1993). Disruption of automatic speech following a right basal ganglia lesion. *Neurology, 43*(9), 1768–1774.

Srisavasdi, R. (March 1, 2007). D.A. drops swearing charges. *Orange County Register*. Retrieved from http://www.ocregister.com/articles/county-59494-law-venable.html.

Steele, C. M. (2010). *Whistling Vivaldi and other clues to how stereotypes affect us*. New York: W. W. Norton and Company.

Stephens, R., Atkins, J., and Kingston, A. (2009). Swearing as a response to pain. *Neuroreport, 20*(12), 1056–1060.

Stork, E., and Hartley, N. T. (2009). Classroom incivilities: Students' perceptions about professors' behaviors. *Contemporary Issues in Education Research, 2*(4), 13.

Taub, S. F. (2001). *Language from the body: Iconicity and metaphor in American Sign Language*. Cambridge: Cambridge University Press.

Telegraph Staff. (January 4, 2010). First words of children include "cat", "beer" and "Hoover." *Telegraph*. Retrieved from http://www.telegraph.co.uk/news/health/children/6929280/First-words-of-children-include-cat-beer-and-Hoover.html.

Thibodeau, P. H., Bromberg, C., Hernandez, R., and Wilson, Z. (2014). An exploratory investigation of word aversion. In *Proceedings of the 36th Annual Conference of the Cognitive Science Society*, P. Bello, M. Guarini, M. McShane, and B. Scassellati (Eds.). Austin, TX: Cognitive Science Society.

Thorndike, E. L. (1920). A constant error in psychological ratings. *Journal of Applied Psychology, 4*, 25–29.

Trask, L. (2004). Where do *mama/papa* words come from? Department of Linguistics and English Language, University of Sussex. Retrieved from https://www.sussex.ac.uk/webteam/gateway/file.php?name=where-do-mama2.pdf&site=1.

TVTropes. (n.d.) Trivia: South Park: Bigger, Longer and Uncut. Retrieved from http://tvtropes.org/pmwiki/pmwiki.php/TriviaSouthParkBigger LongerAndUncut.

Van Lancker, D., and Cummings, J. L. (1999). Expletives: Neurolinguistic and neurobehavioral perspectives on swearing. *Brain Research Reviews, 31*(1), 83–104.

Velten, H. D. V. (1943). The growth of phonemic and lexical patterns in infant language. *Language, 19*(4), 281–292.

Verhulst, B., Lodge, M., and Lavine, H. (2010). The attractiveness halo: Why some candidates are perceived more favorably than others. *Journal of Nonverbal Behavior, 34*(2), 1–2.

von Fleischhacker, R. (1894 [1400]). *Lanfrank's "Science of cirurgie."* (Vol. 1). Periodicals Service Company.

Voutilainen, E. (2008). Kirosanojen kielioppia. *Kotimaisten kielten tutkimuskeskus*. Retrieved from http://www.kotus.fi/nyt/kotuksen_kolumnit/kieli-ikkuna_(1996_2009)/kirosanojen_kielioppia.

Wachal, R. S. (2002). Taboo or not taboo: That is the question. *American Speech*, *77*(2), 195–206.

Wafflesouls. (August 8, 2012). Arrested Development—Chicken Dance (Whole Family). Retrieved from https://www.youtube.com/watch?v =1TphEhoQgvo.

Warner, J. (April 16, 2009). Dude, you've got problems. *New York Times*. Retrieved from http://warner.blogs.nytimes.com/2009/04/16/who-are-you-calling-gay.

Welsh, J. (October 17, 2011). Swearing on TV linked to teen aggression. LiveScience. Retrieved from http://www.livescience.com/16570-profanity-tv-video-games-teen-aggression.html.

Williams, J. N. (1992). Processing polysemous words in context: Evidence for interrelated meanings. *Journal of Psycholinguistic Research*, *21*(3), 193–218.

Wilson, M. D. (1988). The MRC psycholinguistic database: Machine readable dictionary, version 2. *Behavioural Research Methods, Instruments and Computers*, *20*(1), 6–11.

Winslow, A. G. (1900 [1772]). *Diary of Anna Green Winslow: A Boston school girl of 1771*. Boston: Houghton, Mifflin and Company.

Xu, J., Gannon, P. J., Emmorey, K., Smith, J. F., and Braun, A. R. (2009). Symbolic gestures and spoken language are processed by a common neural system. *Proceedings of the National Academy of Sciences*, *106*(49), 20664–20669.

Yung, J., Chang, G. H., and Lai, H. M. (Eds.). (2006). *Chinese American voices: From the Gold Rush to the present*. Berkeley: University of California Press.

Zeitzen, M. (2008). *Polygamy: A cross-cultural analysis*. Oxford: Berg.

Zile, A., and Stephens, R. (2014). Swearing as emotional language. Paper presented at the Annual Conference of the British Psychological Society, Birmingham, UK, May 7–9.

Index

Benjamin K. Bergen is a professor of cognitive science at the University of California, San Diego, where he directs the Language and Cognition Laboratory. His writing has appeared in *Wired*, *Scientific American*, *Psychology Today*, *Salon*, *Time*, *Los Angeles Times*, *Guardian*, and *Huffington Post*. He lives in San Diego.